DEFENSE, SECURITY AND STRATEGIES

NATIONAL NUCLEAR SECURITY ADMINISTRATION AND THE NUCLEAR SECURITY ENTERPRISE

ISSUES AND ASSESSMENTS

DEFENSE, SECURITY AND STRATEGIES

Additional books in this series can be found on Nova's website under the Series tab.

Additional E-books in this series can be found on Nova's website under the E-book tab.

NUCLEAR MATERIALS AND DISASTER RESEARCH

Additional books in this series can be found on Nova's website under the Series tab.

Additional E-books in this series can be found on Nova's website under the E-book tab.

DEFENSE, SECURITY AND STRATEGIES

NATIONAL NUCLEAR SECURITY ADMINISTRATION AND THE NUCLEAR SECURITY ENTERPRISE

ISSUES AND ASSESSMENTS

ALBERT WRIGHT
AND
CRAIG LEWIS
EDITORS

nova publishers
New York

For permission to use material from this book please contact us:
Telephone 631-231-7269; Fax 631-231-8175
Web Site: http://www.novapublishers.com

Library of Congress Cataloging-in-Publication Data

ISBN: 978-1-62808-190-9

Published by Nova Science Publishers, Inc. †New York

CONTENTS

PREFACE

During the late 1990s, the Department of Energy (DOE) had difficulties with a lack of clear management authority and responsibility that contributed to security problems at the nation's nuclear weapons laboratories and management problems with major projects. In response, Congress created the National Nuclear Security Administration (NNSA) as a separately organized agency within the DOE under the NNSA Act. The NNSA is responsible for managing nuclear weapons- and nonproliferation-related national security activities in laboratories and other facilities, collectively known as the nuclear security enterprise. This book examines the problems that have arisen with the nuclear security enterprise, from projects' cost and schedule overruns to inadequate oversight of safety and security at NNSA's sites. With the NNSA proposing to spend tens of billions of dollars to modernize its facilities, it is important to ensure scarce resources are spent in an effective and efficient manner.

Chapter 1 – During the late 1990s, DOE had difficulties with a lack of clear management authority and responsibility that contributed to security problems at the nation's nuclear weapons laboratories and management problems with major projects. In response, Congress created NNSA as a separately organized agency within DOE under the NNSA Act. NNSA is responsible for managing nuclear weapon- and nonproliferation-related national security activities in laboratories and other facilities, collectively known as the nuclear security enterprise. GAO continues to identify problems across the nuclear security enterprise, from projects' cost and schedule overruns to inadequate oversight of safety and security at NNSA's sites. With NNSA proposing to spend tens of billions of dollars to modernize its facilities, it is important to ensure scarce resources are spent in an effective and efficient manner. This testimony addresses (1) NNSA's early experiences organizing and operating as a separately organized agency within DOE and (2) NNSA's efforts to correct long-standing management deficiencies. It is based on prior GAO reports issued from January 1995 to March 2012. DOE and NNSA continue to act on the numerous recommendations GAO has made to improve NNSA's management. GAO will continue to monitor DOE's and NNSA's implementation of these recommendations.

Chapter 2 – NNSA has primary responsibility for ensuring the safety, security, and reliability of the nation's nuclear weapons stockpile. NNSA carries out these activities at three national labs, four production sites, and one test site—collectively known as the nuclear security enterprise. Contractors operate these sites under management and operations (M&O) contracts. The enterprise workforces often possess certain critical skills that can only be developed through a minimum of 3 years of experience working in a secure, classified environment. Because NNSA could have difficulty maintaining the critically skilled workforces necessary to ensure the safety, security, and reliability of the nation's nuclear weapons, GAO was asked to examine: (1) strategies NNSA and its M&O contractors use to recruit, develop, and retain critically skilled workforces; (2) how NNSA assesses the effectiveness of these strategies; and (3) challenges in recruiting, retaining, and developing this specialized workforce and efforts to mitigate these challenges. GAO reviewed NNSA's and its M&O contractors' human capital documents and interviewed officials.

Chapter 3 – NNSA, a semiautonomous agency within DOE, is responsible for the nation's nuclear weapons, nonproliferation, and naval reactors programs. Since its inception in 2000, the agency has faced challenges in its ability to accurately identify the costs of major projects. In addition, both the DOE Inspector General, in 2003, and GAO, in 2007, reported concerns with NNSA's PPBE process, specifically in how NNSA validates budget estimates and decides on resource allocations or trade-offs. GAO was asked to review how NNSA manages programming and budgeting through its PPBE process. GAO examined (1) the current structure of NNSA's PPBE process, (2) the extent to which NNSA reviews its budget estimates, and (3) how NNSA decides on resource trade-offs in its PPBE process. To carry out its work, GAO reviewed NNSA policies, instructions, guidance, and internal reports documenting the agency's PPBE process and interviewed NNSA, DOE, and M&O contractor officials.

Chapter 4 – DOE and NNSA are responsible for managing nuclear weapon- and nonproliferation-related national security activities in national laboratories and other sites and facilities, collectively known as the nuclear security enterprise. Major portions of NNSA's mission are largely carried out by contractors at each site. GAO has designated contract management of major projects (i.e., those $750 million or more) at DOE and NNSA as a high risk area. Progress has been made, but GAO continues to identify security and safety problems at DOE and NNSA sites as well as project and contract management problems related to cost and schedule overruns on major projects. This testimony addresses DOE's and NNSA's oversight of (1) security performance, (2) safety performance, and (3) project and contract management in the nuclear security enterprise. It is based on prior GAO reports issued from August 2000 to December 2012. DOE and NNSA continue to act on the numerous recommendations GAO has made to improve management of the nuclear security enterprise. GAO will continue to monitor DOE's and NNSA's implementation of these recommendations.

In: National Nuclear Security Administration ... ISBN: 978-1-62808-190-9
Editors: A. Wright and C. Lewis

Chapter 1

MODERNIZING THE NUCLEAR SECURITY ENTERPRISE: OBSERVATIONS ON THE ORGANIZATION AND MANAGEMENT OF THE NATIONAL NUCLEAR SECURITY ADMINISTRATION*

Gene Aloise, U.S. Government Accountability Office

WHY GAO DID THIS STUDY

During the late 1990s, DOE had difficulties with a lack of clear management authority and responsibility that contributed to security problems at the nation's nuclear weapons laboratories and management problems with major projects. In response, Congress created NNSA as a separately organized agency within DOE under the NNSA Act. NNSA is responsible for managing nuclear weapon- and nonproliferation-related national security activities in laboratories and other facilities, collectively known as the nuclear security enterprise. GAO continues to identify problems across the nuclear security enterprise, from projects' cost and schedule overruns to inadequate oversight of safety and security at NNSA's sites. With NNSA proposing to spend tens of

* This is an edited, reformatted and augmented version of The United States Government Accountability Office publication, GAO-12-867T, dated June 27, 2012.

billions of dollars to modernize its facilities, it is important to ensure scarce resources are spent in an effective and efficient manner.

This testimony addresses (1) NNSA's early experiences organizing and operating as a separately organized agency within DOE and (2) NNSA's efforts to correct long-standing management deficiencies. It is based on prior GAO reports issued from January 1995 to March 2012.

DOE and NNSA continue to act on the numerous recommendations GAO has made to improve NNSA's management. GAO will continue to monitor DOE's and NNSA's implementation of these recommendations.

What GAO Found

After the enactment of Title 32 of the National Defense Authorization Act for Fiscal Year 2000 (NNSA Act), the Department of Energy (DOE) and the National Nuclear Security Administration (NNSA) struggled to determine how NNSA should operate as a separately organized agency within the department. A number of factors contributed to this. First, DOE and NNSA did not have a useful model to follow for establishing a separately organized agency in DOE. Several federal agencies were suggested as models, such as the National Oceanic and Atmospheric Administration in the Department of Commerce. However, GAO reported in January 2007 that agency officials GAO interviewed did not consider their agency to be separately organized or believed that their agency's operational methods were transferable to NNSA. Second, DOE's January 2000 plan to implement the NNSA Act did not define how NNSA would operate as a separately organized agency within DOE. Internal DOE opposition to the creation of NNSA led the department to fill virtually every significant statutory position in NNSA with DOE officials (i.e., having DOE officials contemporaneously serve in NNSA and DOE positions). As GAO testified in April 2001, this practice of "dualhatting" caused considerable concern about NNSA's ability to independently function. Also, lack of formal agreement between DOE and NNSA in a number of key areas such as, among others, budgeting and procurement, led to organizational conflicts that inhibited effective operations. Even where formal procedures were developed, interpersonal disagreements hindered effective cooperation. For example, a January 2007 GAO report described the conflict between NNSA and DOE counterintelligence offices, which led to Congress subsequently amending the NNSA Act to consolidate the counterintelligence programs of DOE and NNSA under DOE.

NNSA has made considerable progress resolving some of its long-standing management deficiencies, but significant improvement is still needed especially in NNSA's management of its major projects and contracts. GAO reported in June 2004 that NNSA has better delineated lines of authority and has improved communication between its headquarters and site offices. In addition, NNSA's establishment of an effective headquarters security organization has made significant progress resolving many of the security weaknesses GAO has identified. Nevertheless, NNSA continues to experience major cost and schedule overruns on its projects, such as research and production facilities and nuclear weapons refurbishments, principally because of ineffective oversight and poor contractor management. In some areas, NNSA can be viewed as a success. Importantly, NNSA has continued to ensure that the nuclear weapons stockpile remains safe and reliable in the absence of underground nuclear testing. At the same time, NNSA's struggles in defining itself as a separately organized agency within DOE, and the considerable management problems that remain have led to calls in Congress and other organizations to increase NNSA's independence from DOE. However, senior DOE and NNSA officials have committed to continuing reform, and DOE's and NNSA's efforts have led to some management improvements. As a result, GAO continues to believe, as it concluded in its January 2007 report, that drastic organizational change to increase independence is unnecessary and questions whether such change would solve the agency's remaining management problems.

Chairman Turner, Ranking Member Sanchez, and Members of the Subcommittee:

We are pleased to be here today to discuss the creation and implementation of the National Nuclear Security Administration (NNSA)— a separately organized agency within the Department of Energy (DOE). As you know, NNSA is responsible for the management and security of the nation's nuclear weapons, nuclear nonproliferation, and naval reactor programs at research and development laboratories, production plants, and other facilities known collectively as the nuclear security enterprise.[1]

During the late 1990s, DOE experienced management difficulties with its nuclear weapons program that contributed to security problems at the nation's nuclear weapons laboratories and significant cost overruns on major projects. According to a June 1999 report by the President's Foreign Intelligence Advisory Board, DOE's management of the nuclear weapons laboratories, while representing "science at its best," also embodied "security at its worst" because of "organizational disarray, managerial neglect, and...a culture of arrogance." The board urged Congress to create a new organization that, whether established as an independent agency or a semiautonomous agency within DOE, would have a clear mission, streamlined bureaucracy, and drastically simplified lines of authority and accountability. Responding to the board's recommendations, Congress created NNSA under Title 32 of the National Defense Authorization Act for Fiscal Year 2000—the NNSA Act.[2]

The NNSA Act established NNSA as a "separately organized agency" within DOE. The act established the position of DOE Under Secretary for Nuclear Security, who was also designated as the Administrator of NNSA. The Secretary of Energy and the Deputy Secretary of Energy were allowed to establish policy for NNSA and to give direction to NNSA through the Administrator; however, other DOE employees were prohibited from directing the activities of individual NNSA employees. In addition, the NNSA Act required that, among other things, NNSA develop a planning, programming, and budgeting process to ensure that NNSA operated under sound financial management principles. Using this planning, programming, and budgeting process, NNSA is also required to annually submit to Congress a Future Years Nuclear Security Program (FYNSP) plan that details NNSA's planned expenditures for the next 5 years.

DOE's and NNSA's management of the nuclear security enterprise has been the subject of much criticism. The department's problems are long-standing. For example, we first designated DOE's management of its contracts as an area at high risk of fraud, waste, abuse, and mismanagement in 1990 because of the department's record of inadequate management and oversight of its contractors. In January 1995, we reported that DOE's laboratories did not have clearly defined missions that focus their considerable resources on accomplishing the department's changing objectives and national priorities.[3] Noting that the laboratories have made vital contributions to the nation's defense and civilian science and technology efforts, we reported that DOE had not coordinated these laboratories' efforts to solve national problems but had instead managed each laboratory on a program-by-program basis. The establishment of NNSA as a semiautonomous agency within DOE in 2000 was intended to correct these long-standing and widely recognized DOE management problems, which had been underscored by significant cost overruns on major projects and security problems at the national laboratories.

NNSA's creation, however, has not yet had the desired effect of fully resolving these management problems. Progress has been made, but NNSA and DOE's Office of Environmental Management remain on our high-risk list. 4 Furthermore, we continue to identify problems across the nuclear security enterprise, ranging from significant cost and schedule overruns on major projects to ineffective federal oversight of safety and security at NNSA's sites.[5] Concerns have also been raised by national laboratory and other officials that DOE's and NNSA's oversight of the laboratories' activities has been excessive and that the safety and security requirements the laboratories' are subject to are overly prescriptive and burdensome, which has resulted in a negative effect on the quality of science performed at these laboratories.

In January 2007, we testified before this Subcommittee on the extent to which NNSA has taken steps to improve security at its facilities, improve its management practices, and revise its organizational structure. 6 Similarly, in February 2012, we testified before this Subcommittee on NNSA's management of the nuclear security enterprise.[7] My testimony today, which is based on these and other reports and testimonies we have issued since NNSA's creation, discusses (1) NNSA's early experiences organizing and operating as a separately organized agency within DOE and (2) NNSA's efforts to correct long-standing management deficiencies. Detailed information about scope and methodology can be found in our issued reports. We conducted the performance audit work that supports this statement in accordance with generally accepted government auditing standards. Those standards require that we plan and perform audits to obtain sufficient, appropriate evidence to provide a reasonable basis for our findings and conclusions based on our audit objectives. We believe that the evidence obtained provides a reasonable basis for our findings and conclusions based on our audit objectives.

BACKGROUND

DOE is responsible for a diverse set of missions, including nuclear security, energy research, and environmental cleanup. These missions are managed by various organizations within DOE and largely carried out by contractors at DOE sites. According to federal budget data, NNSA is the largest organization in DOE, overseeing nuclear weapons, nuclear nonproliferation, and naval reactors missions at its sites. With a $10.5 billion budget in fiscal year 2011—nearly 40 percent of DOE's total budget—NNSA is responsible for, among other things, providing the United States with safe, secure, and reliable nuclear weapons in the absence of underground nuclear testing and maintaining core competencies in nuclear weapons science, technology, and engineering. Ensuring that the nuclear weapons stockpile remains safe and reliable in the absence of underground nuclear testing is extraordinarily complicated and requires state-of-the-art experimental and computing facilities, as well as the skills of top scientists in the field. Over the past decade, the United States has invested billions of dollars in sustaining the cold war-era stockpile and upgrading the laboratories and, in 2011, the administration announced plans to request $88 billion from Congress over the next decade to operate and modernize the nuclear security enterprise and ensure that base scientific, technical, and engineering capabilities are sufficiently supported, and the nuclear deterrent in the United States can continue to be safe, secure, and reliable.

Under DOE's long-standing model of having unique management and operating (M&O) contractors at each site, management of its sites has historically been decentralized and, thus, fragmented. Since the Manhattan Project produced the first atomic bomb during World War II, NNSA, DOE, and their predecessor agencies have depended on the expertise of private firms, universities, and others to carry out research and development work and efficiently operate the facilities necessary for the nation's nuclear defense. DOE's relationship with these entities has been formalized over the years through its M&O contracts—agreements that give DOE's contractors unique responsibility to carry out major portions of DOE's missions and apply their scientific, technical, and management expertise.[8]

Currently, DOE spends 90 percent of its annual budget on M&O contracts, making it the largest non-Department of Defense contracting agency in the government. The M&O contractors at DOE's NNSA sites have operated under DOE's direction and oversight but largely independently of one another. Various headquarters and field-based organizations within DOE and NNSA develop policies, and NNSA site offices, collocated with NNSA's sites, conduct day-to-day oversight of the M&O contractors and evaluate the contractors' performance in carrying out the sites' missions.

DOE and NNSA Struggled to Determine How NNSA Should Operate as a Separately Organized Agency

NNSA focused considerable attention on reorganizing its internal operations; however, it and DOE have struggled with establishing how NNSA should operate as a separately organized agency within the department. Several factors contributed to this situation. First, DOE and NNSA did not have a useful model to follow for establishing a separately organized agency in DOE. The President's Foreign Intelligence Advisory Board's June 1999 report suggested several federal agencies, such as the National Oceanic and Atmospheric Administration in the Department of Commerce, which could be used as a model for NNSA. However, as we reported in January 2007, none of the agency officials we interviewed considered their agency to be separately organized or believed that their agency's operational methods were transferable to NNSA.[9] Second, DOE's January 2000 implementation plan, which was required by the NNSA Act, did not define how NNSA would operate as a separately organized agency within DOE. Instead, reflecting the opposition of the then DOE senior leadership to the creation of NNSA, the implementation plan "dual-hatted" virtually every significant statutory position in NNSA with DOE officials (i.e., having DOE officials contemporaneously serve in NNSA and DOE positions), including the Director of NNSA's Office of Defense Nuclear Counterintelligence and General Counsel. As we testified in April 2001, this practice caused considerable concern about NNSA's ability to function with the independence envisioned in the NNSA Act.[10] Dual-hatting was subsequently forbidden by an amendment to the NNSA Act.[11]

A lack of formal agreement between DOE and NNSA in a number of key areas—budgeting, procurement, information technology, management and administration, and safeguards and security—resulted in organizational conflicts that inhibited effective operations. Even where formal procedures were developed, interpersonal disagreements hindered effective cooperation. For example, our January 2007 report described the conflict between NNSA and DOE counterintelligence offices.[12] Specifically, NNSA and DOE counterintelligence officials disagreed over (1) the scope and direction of the counterintelligence program, (2) their ability to jointly direct staff in the headquarters counterintelligence program offices, (3) the allocation of counterintelligence resources, (4) counterintelligence policy making and (5) their roles and responsibilities in handling specific counterintelligence matters. Subsequently, Congress amended the NNSA Act to consolidate the counterintelligence programs of DOE and NNSA under the Department of Energy.[13]

These persistent challenges defining NNSA's role as a separately organized agency have led to calls in Congress and other organizations to enhance NNSA's ability to operate independently of DOE. For example, the

Defense Science Board proposed in 2006 that a completely independent nuclear weapons agency be created. [14] DOE's Office of Inspector General has also recently questioned the relationship between DOE and NNSA. Specifically, in November 2011, DOE's Office of Inspector General reported that NNSA, as a result of its separately organized status, maintains a costly set of distinctly separate overhead and indirect cost operations that often duplicate existing DOE functions.[15] For example, NNSA retains separate functions in areas such as, among others, congressional affairs, general counsel, human resources, procurement and acquisition, and public affairs. According to this November 2011 report, these redundant operations are costly and can complicate communications and program execution. There have been continuing calls for removing NNSA from DOE and establishing it as a separate agency. We reported in January 2007 that former senior DOE and NNSA officials with whom we spoke generally did not favor removing NNSA from DOE; we concluded that such drastic change was unnecessary to produce an effective organization.[16]

NNSA Has Made Considerable Improvements, but Deficiencies Persist, Especially in Management of Major Projects and Contracts

Since its creation, NNSA has made considerable progress resolving some of its long-standing management deficiencies. For example, we reported in June 2004 that NNSA had better delineated lines of authority and improved communication between NNSA headquarters and its site offices.[17] Furthermore, our January 2007 report contained 21 recommendations to the Secretary of Energy and the Administrator of NNSA that were intended to correct deficiencies in five areas— organization, security, project management, program management, and financial management. DOE and NNSA have taken important steps to address most of these recommendations. For example, to improve security, we recommended that the Administrator of NNSA, among other things, implement a professional development program for security staff to ensure the completion of needed training, develop a framework to evaluate results from security reviews and guide security improvements, and establish formal mechanisms for sharing and implementing lessons learned across the weapons complex. NNSA's establishment of an effective headquarters security organization has made significant progress implementing these recommendations by performing security reviews, developing security performance measures, and instituting a security lessons-learned center.

Nevertheless, NNSA continues to experience significant deficiencies, particularly in its management of major projects and contracts. As we testified in February 2012, a basic tenet of effective management is the ability to complete projects on time and within budget.[18] However, for more than a decade, NNSA has continued to experience significant cost and schedule overruns on its major projects, principally because of ineffective oversight and poor contractor management. We have reported that NNSA's efforts to extend the operational lives of nuclear weapons in the stockpile have experienced cost increases and schedule delays, such as a $300 million cost increase and 2-year delay in the refurbishment of the W87 nuclear warhead and a $70 million cost increase and 1-year delay in the refurbishment of the W76 nuclear warhead.[19] Furthermore, we reported that the estimated cost to construct a modern Uranium Processing Facility at NNSA's Y-12 National Security Complex experienced a nearly sevenfold cost increase from between $600 million and $1.1 billion in 2004 to between $4.2 billion and $6.5 billion in 2011.[20] We also reported in March 2012 that NNSA's project to construct a new plutonium research facility at Los Alamos National Laboratory—the Chemistry and Metallurgy Research Replacement Nuclear Facility— would cost between $3.7 billion and $5.8 billion—nearly a sixfold increase from NNSA's original estimate.[21] NNSA's February 2012 decision to defer construction of this facility for at least 5 years will result in a total delay of between 8 and 12 years from its original plans.

NNSA's planning, programming, and budgeting process has also experienced a setback, which raises questions about the process's capability and flexibility. Specifically, NNSA's modernization and operations plans are detailed and annually updated in the agency's Stockpile Stewardship and Management Plan (SSMP), which provides details of nuclear security enterprise modernization and operations plans over the next two decades. In addition, as discussed above, the NNSA Act requires NNSA to annually submit to Congress an FYNSP—a budget document approved by the Office of Management and Budget that details NNSA's planned expenditures for the next 5 years. Furthermore, Section 1043 of the National Defense Authorization Act for Fiscal Year 2012 requires the Department of Defense and NNSA to jointly produce an annual report that, among other things, provides a detailed 10-year estimate of modernization budget requirements. NNSA neither submitted an FYNSP based on "programmatic requirements"[22] nor the Section 1043 annual report with its fiscal year 2013 budget submission. In addition, NNSA has yet to release an updated SSMP. According to the Secretary of Energy, the August 2011 Budget Control Act created "new fiscal realities" that have caused the agency to revise its long-range modernization and operations plans and budget.[23] An NNSA official told us that the revised plans, which will include the FYNSP, Section 1043 annual report, and updated SSMP should be completed in July 2012. We are currently reviewing NNSA's planning, programming, and budgeting process in response to a request from the Subcommittee on Energy and Water Development, Senate Committee on Appropriations, and we expect to issue a report on this work in the next few months.

In conclusion, producing a well-organized and effective agency out of what was widely considered a dysfunctional enterprise has been a considerable challenge. In some areas, NNSA can be viewed as a success. In particular, NNSA has successfully ensured that the nuclear weapons stockpile remains safe and reliable in the absence of underground nuclear testing, accomplishing this complicated task by using state-of-the-art facilities, as well as the skills of top scientists. As we testified in February 2012, maintaining government-owned facilities that were constructed more than 50 years ago and ensuring M&O contractors are sustaining critical human capital skills that are highly technical in nature and limited in supply are both difficult undertakings. Careful federal oversight over the tens of billions of dollars NNSA proposes to spend to modernize nuclear facilities will be necessary to ensure these funds are spent in as an effective and efficient manner as possible, especially given NNSA's record of weak management of its major projects.

Over the past decade, we have made numerous recommendations to DOE and NNSA to improve their management and oversight practices. DOE and NNSA have acted on many of these recommendations and have made considerable progress. Nevertheless, enough significant management problems remain that prompt some to call for removing NNSA from DOE and either moving it to another department or establishing it as a separate agency. As we concluded in January 2007, however, we do not believe that such drastic changes are necessary, and we continue to hold this view today. Importantly, we are uncertain whether such significant organizational changes to increase NNSA's independence would produce the desired effect of creating a modern, responsive, effective, and efficient nuclear security enterprise. In light of the substantial leadership commitment to reform made by senior DOE and NNSA officials, and the significant improvements that have already been made, we believe that NNSA remains capable of delivering the management improvements necessary to be an effective organization, and we will continue to monitor NNSA's progress making these improvements.

Chairman Turner, Ranking Member Sanchez, and Members of the Subcommittee, this completes my prepared statement. I would be pleased to respond to any questions you may have at this time.

End Notes

[1] Specifically, NNSA manages three national nuclear weapon design laboratories— Lawrence Livermore National Laboratory in California, Los Alamos National Laboratory in New Mexico, and Sandia National Laboratories in New Mexico and California. It also manages four nuclear weapons production plants—the Pantex Plant in Texas, the Y-12 National Security Complex in Tennessee, the Kansas City Plant in Missouri, and the Tritium Extraction Facility at DOE's Savannah River Site in South Carolina. NNSA also manages the Nevada National Security Site, formerly known as the Nevada Test Site.

[2] Pub. L. No. 106-65, 113 Stat. 512, 953 (1999).

[3] GAO, Department of Energy: National Laboratories Need Clearer Missions and Better Management, GAO/RCED-95-10 (Washington, D.C.: Jan. 27, 1995).

[4] GAO, High-Risk Series: An Update, GAO-11-278 (Washington, D.C.: February 2011).

[5] GAO, Department of Energy: Views on the Progress of the National Nuclear Security Administration in Implementing Title 32, GAO-01-602T (Washington, D.C.: Apr. 4, 2001); GAO, NNSA Management: Progress in the Implementation of Title 32, GAO-02-93R (Washington, D.C.: Dec. 12, 2001); and GAO, Department of Energy: NNSA Restructuring and Progress in Implementing Title 32, GAO-02-451T (Washington, D.C.: Feb. 26, 2002).

[6] GAO, National Nuclear Security Administration: Security and Management Improvements Can Enhance Implementation of the NNSA Act, GAO-07-428T (Washington, D.C.: Jan. 31, 2007).

[7] GAO, National Nuclear Security Administration: Observations on NNSA's Management and Oversight of the Nuclear Security Enterprise, GAO-12-473T (Washington, D.C.: Feb. 16, 2012).

[8] M&O contracts are agreements under which the government contracts for the operation, maintenance, or support, on its behalf, of a government-owned or -controlled research, development, special production, or testing establishment wholly or principally devoted to one or more of the major programs of the contracting federal agency. Federal Acquisition Regulation, 48 C.F.R. § 17.601.

[9] GAO, National Nuclear Security Administration: Additional Actions Needed to Improve Management of the Nation's Nuclear Programs, GAO-07-36 (Washington, D.C.: Jan. 19, 2007). We interviewed agency officials from the Department of Commerce's National Oceanic and Atmospheric Administration, the Defense Advanced Research Projects Agency, the Defense Threat Reduction Agency, and the Department of Transportation's Federal Aviation Administration.

[10] GAO-01-602T

[11] Pub. L. 106-398, § 3157 (2000) (codified at 50 U.S.C. § 2410).

[12] GAO-07-36.

[13] Section 3117 of the John Warner National Defense Authorization Act for Fiscal Year 2007 contained provisions to temporarily consolidate the counterintelligence programs of DOE and NNSA under the Department of Energy. Pub. L. No. 109-364, § 3117 (2006). In 2009, Congress made this consolidation permanent. Pub. L. No. 111-84, § 3121 (2009).

[14] The Defense Science Board provides the Department of Defense with independent advice and recommendations on matters relating to the department's scientific and technical enterprise See Defense Science Board Task Force, Nuclear Capabilities (Washington, D.C.: December 2006).

[15] DOE Office of Inspector General, Special Report: Management Challenges at the Department of Energy, DOE/IG-0858 (Washington, D.C.: November 2011).

[16] GAO-07-36

[17] GAO, National Nuclear Security Administration: Key Management Structure and Workforce Planning Issues Remain as NNSA Conducts Downsizing, GAO-04-545 (Washington, D.C.: June 25, 2004).

[18] GAO-12-473T.

[19] GAO, Nuclear Weapons: Improved Management Needed to Implement Stockpile Stewardship Program Effectively, GAO-01-48 (Washington, D.C.: Dec. 14, 2000) and GAO, Nuclear Weapons: NNSA and DOD Need to More Effectively Manage the Stockpile Life Extension Program, GAO-09-385 (Washington, D.C.: Mar. 2, 2009).

[20] GAO, Nuclear Weapons: National Nuclear Security Administration's Plans for Its Uranium Processing Facility Should Better Reflect Funding Estimates and Technology Readiness, GAO-11-103 (Washington, D.C.: Nov. 19, 2010).

[21] GAO, Modernizing the Nuclear Security Enterprise: New Plutonium Research Facility at Los Alamos May Not Meet All Mission Needs, GAO-12-337 (Washington, D.C.: Mar. 26, 2012).

[22] The NNSA fiscal year 2013 budget submission said that future year funding levels based on actual programmatic requirements will be produced a later date.

[23] The Budget Control Act of 2011, amending the Balanced Budget and Emergency Deficit Control Act of 1985, establishes limits on discretionary spending for fiscal years 2012 through 2021. In addition, the act specifies additional limits on discretionary spending and automatic reductions in direct spending because legislation was not enacted that would reduce projected deficits by at least $1.2 trillion by the end of fiscal year 2021. Among other things, the Budget Control Act requires the Office of Management and Budget to calculate, and the President to order, a sequestration of discretionary and direct spending on January 2, 2013, to achieve reductions for that fiscal year. See GAO, Agency Operations: Agencies Must Continue to Comply with Fiscal Laws Despite the Possibility of Sequestration, GAO-12-675T (Washington, D.C.: Apr. 25, 2012).

In: National Nuclear Security Administration ... ISBN: 978-1-62808-190-9
Editors: A. Wright and C. Lewis

Chapter 2

MODERNIZING THE NUCLEAR SECURITY ENTERPRISE: STRATEGIES AND CHALLENGES IN SUSTAINING CRITICAL SKILLS IN FEDERAL AND CONTRACTOR WORKFORCES*

United States Government Accountability Office

WHY GAO DID THIS STUDY

NNSA has primary responsibility for ensuring the safety, security, and reliability of the nation's nuclear weapons stockpile. NNSA carries out these activities at three national labs, four production sites, and one test site—collectively known as the nuclear security enterprise. Contractors operate these sites under management and operations (M&O) contracts. The enterprise workforces often possess certain critical skills that can only be developed through a minimum of 3 years of experience working in a secure, classified environment.

---

* This is an edited, reformatted and augmented version of the Highlights of GAO-12-468, a report to the Subcommittee on Oversight of Government Management, the Federal Workforce, and the District of Columbia, Committee on Homeland Security and Governmental Affairs, U.S. Senate, dated April 2012.

Because NNSA could have difficulty maintaining the critically skilled workforces necessary to ensure the safety, security, and reliability of the nation's nuclear weapons, GAO was asked to examine: (1) strategies NNSA and its M&O contractors use to recruit, develop, and retain critically skilled workforces; (2) how NNSA assesses the effectiveness of these strategies; and (3) challenges in recruiting, retaining, and developing this specialized workforce and efforts to mitigate these challenges. GAO reviewed NNSA's and its M&O contractors' human capital documents and interviewed officials.

WHAT GAO RECOMMENDS

GAO recommends that NNSA consider developing standardized definitions for human capital metrics across the enterprise to ensure NNSA and its M&O contractors gather consistent contractor data. NNSA concurred with GAO's recommendation.

WHAT GAO FOUND

The National Nuclear Security Administration (NNSA) and its M&O contractors have developed and implemented multifaceted strategies to recruit, develop, and retain both the federal and contractor workforces needed to preserve critical skills in the enterprise. NNSA's recruiting and retention efforts for its federal staff focus on attracting early career hires with competitive pay and development opportunities. Its development efforts generally rely on two key programs to develop its critically skilled workforce––one that identifies needs and another that identifies the qualifications necessary to meet them. For strategic planning purposes, NNSA is also undertaking a comprehensive reassessment and analysis of staffing requirements to ascertain future federal workforce requirements. M&O contractors' recruitment and retention strategies vary from site to site, but each site focuses on maintaining competitive compensation packages. Their development efforts vary in approach and scope and face some challenges—particularly in preserving underground nuclear testing skills.

To assess the effectiveness of its own—and its M&O contractors'—strategies for recruiting, developing, and retaining the workforces needed to preserve critical skills, NNSA monitors key human capital metrics. NNSA focuses on two key metrics in assessing its own strategies—the time it takes to hire a new employee and its attrition rates. To assess the effectiveness of its contractors' strategies, NNSA monitors key human capital metrics using data that M&O contractors collect, including acceptance rates, attrition rates, comparability of pay and benefits with peer institutions, and the ability to fill a critical skills position within a certain number of days. M&O contractors assess key human capital performance measures, but these metrics do not have standardized definitions. For example, one of the M&O contractors' key metrics—acceptance rates for offers of employment—may not be consistently measured across the enterprise. Without this information, NNSA's ability to monitor the effectiveness of its and its M&O contractors' strategies to recruit, develop, and retain the workforces needed to preserve critical skills may be hindered. In particular, without common enterprisewide definitions of human capital performance metrics, NNSA may not be able to collect consistent and comparable data across all eight sites in the enterprise.

The enterprise's work environments and site locations pose recruiting challenges, and NNSA and its M&O contractors face shortages of qualified candidates, among other challenges. For example, staff must often work in secure areas that prohibit the use of personal cell phones, e-mail, and social media, which is a disadvantage in attracting younger skilled candidates. In addition, many sites are geographically isolated and may offer limited career opportunities for candidates' spouses. Critically skilled positions also require security clearances—and therefore U.S. citizenship—and a large percentage of students graduating from top science, technology, and engineering programs are foreign nationals. The pool of qualified candidates is also attractive to high technology firms in the private sector, which may offer more desirable work environments. NNSA and its M&O contractors are taking actions to address these challenges where possible, including streamlining hiring and security clearance processes and taking actions to proactively identify new scientists and engineers to build a pipeline of critically skilled candidates.

ABBREVIATIONS

CHRD	Contractor Human Resources Division
DOE	Department of Energy
KPP	Knowledge Preservation Program
M&O	management and operations

NNSA National Nuclear Security Administration
TQP Technical Qualification Program

April 26, 2012

The Honorable Daniel K. Akaka
Chairman
The Honorable Ron Johnson
Ranking Member
Subcommittee on Oversight of Government Management, the Federal Workforce, and the District of Columbia
Committee on Homeland Security and Governmental Affairs
United States Senate

The National Nuclear Security Administration (NNSA)—a separately organized agency within the Department of Energy (DOE)—has primary responsibility for ensuring the safety, security, and reliability of the nation's nuclear weapons stockpile.[1] NNSA carries out these activities at eight government-owned, contractor-operated sites, which include three national laboratories, four production plants, and one test site. Collectively, these sites are referred to as the nuclear security enterprise. The enterprise, formerly known as the nuclear weapons complex, has been a significant component of U.S. national security since the 1940s. Contractors operate sites within the enterprise under management and operations (M&O) contracts.[2] These contracts provide the contractor with broad discretion in carrying out the mission of the particular contract but grant the government the option to become much more directly involved in day-to-day management and operations.

Historically, confidence in the safety and reliability of the nuclear stockpile was derived through a continuous process of designing, testing, and deploying new weapons to replace older weapons. In 1992, at the end of the Cold War, and in response to a congressionally imposed U.S. nuclear test moratorium,[3] the United States ceased underground testing of nuclear weapons, and adopted the Stockpile Stewardship Program as an alternative to testing and producing new weapons. The Stockpile Stewardship Program primarily relies on analytical simulations and computer modeling to make expert judgments about the safety, security, and reliability of the nation's nuclear weapons. In addition, NNSA refurbishes weapons in the stockpile to extend their operational lives. Under current national policy, NNSA may also be called upon to resume underground nuclear testing at the Nevada National Security Site within a 3-year time frame under certain circumstances, including the accumulation of uncertainties about the reliability of the nuclear stockpile.

Currently, NNSA's workforce is made up of about 34,000 M&O contractor employees that span the enterprise, and about 2,400 federal employees directly employed by NNSA in its Washington headquarters, at site offices located at each of the eight enterprise sites, and at its Albuquerque, New Mexico, complex. NNSA's staff provide leadership and program management for the nuclear security enterprise and support and oversee its M&O contractors by providing business, technical, financial, legal, and management advice, including support for contractor workforce planning and restructuring, compensation, benefits, oversight of labor- management relations, and the quality of contractor deliverables such as nuclear weapons components. Many workers in the enterprise—both NNSA's staff and its M&O contractors—possess certain critical skills not readily available in the job market. These workers often have advanced degrees in scientific or engineering fields or experience in high-skill, advanced manufacturing techniques. In addition, certain critical skills are unique to the enterprise and, according to NNSA officials, can only be developed within its secure, classified environment. According to these officials, it generally takes a minimum of 3 years of on-the-job training to achieve the skills necessary to succeed in most critical skills positions. Some nuclear weapons expertise can take even longer to develop and must be gained through several years of mentoring, training, and on-the- job experience. For example, according to officials at Los Alamos National Laboratory, it takes 5 to10 years to train a scientist or engineer with an advanced degree to be a fully qualified nuclear weaponeer.

Over the last 20 years, in an effort to operate more efficiently and at reduced cost, DOE has sharply reduced its enterprise contractor workforce—from approximately 52,000 in 1992 to its current level of about 34,000. This decrease raised concerns about preserving critical skills in the enterprise. In 1999, a report from a congressionally mandated commission warned that unless DOE acted quickly to recruit and retain its critically skilled staff and M&O contractor employees—and sharpen the expertise already available—the department could have difficulty ensuring the safety, security, and reliability of the nation's nuclear weapons.[4] DOE, and later NNSA, took steps to correct these problems, and in February 2005, we reported that these efforts had been generally effective.[5] However, in February 2011, in a report assessing the extent to which NNSA has the data necessary to make informed, enterprisewide decisions,[6] we found that NNSA did not have comprehensive information on the status of its M&O contractor workforce. In particular, we reported that NNSA did not have data on the critical skills needed to maintain the Stockpile Stewardship Program's capabilities. As a result, we recommended that NNSA establish a plan with time frames and milestones for the development of a comprehensive contractor workforce baseline that includes the identification of critical human capital skills, competencies, and levels needed to maintain the nation's nuclear weapons strategy. NNSA stated that it understood all of our recommendations in that report and believed that it could implement them. As of March 2012, NNSA had completed a draft plan and was incorporating stakeholders' comments. NNSA officials said that they expect to complete the final contractor workforce baseline plan by May 2012.

NNSA expressed concerns in its FY 2012 *Stockpile Stewardship Management Plan* about the state of both its federal and contractor workforces, stating that there was an urgent need to "refresh" both. In particular, NNSA noted that many employees have retired or are expected to retire soon. At the same time, NNSA's mission has become even more dependent on high-level science, computer science, technology, and engineering skills as it has moved from underground testing as a means for assessing the safety and reliability of nuclear weapons to one dependent on advanced computer simulations, analyses, and nonnuclear tests. These changes make it even more important that NNSA and its M&O contractors preserve critical skills in their workforces. Additional concerns about human capital in the enterprise have been raised by the debate over—and eventual ratification of—the New Start Treaty,[7] which commits the United States to reduce the size of its strategic nuclear weapon stockpile from a maximum of 2,200 to 1,550 nuclear weapons. Reductions in the number of nuclear weapons make it all the more important that NNSA and contractor staff have the requisite critical skills to maintain the safety, security, and reliability of the remaining weapons. However, as the enterprise has contracted, NNSA officials note that training opportunities have been limited, leaving little or no redundancy in certain critical skills within the contractor workforce.

In this context, you asked us to examine NNSA's human capital planning. Specifically, our objectives were to examine: (1) the strategies NNSA and its M&O contractors use to recruit, develop, and retain the workforces needed to preserve the critical skills in the enterprise; (2) how NNSA assesses the effectiveness of these strategies; and (3) challenges that NNSA and its M&O contractors face in recruiting, retaining, and developing this specialized workforce and their efforts to mitigate these challenges.

To address these three objectives, we conducted interviews with human capital planning officials at NNSA headquarters, the Albuquerque complex in New Mexico, and all eight NNSA site offices. We also obtained and reviewed NNSA information about recruiting and retention practices for critically skilled employees, as well as each site's efforts to preserve knowledge needed to sustain critical capabilities. We visited six of the eight sites in the enterprise, including the three national laboratories, Los Alamos National Laboratory and Sandia National Laboratories in New Mexico and Lawrence Livermore National Laboratory in California; two of the production plants, the Pantex Plant in Texas and the Y-12 Plant in Tennessee; and the test site, Nevada National Security Site in Nevada. We conducted telephone interviews with human capital managers at the two other production plants, the Kansas City Plant in Missouri and the Savannah River Site in South Carolina. To examine the strategies NNSA and its M&O contractors use to recruit and retain critically skilled workers, we collected key workforce data from each facility, including NNSA and M&O contractor reports and other documents on the performance and progress made in meeting recruitment and retention targets. To identify challenges in retaining, recruiting, and developing the critical skills workforce, we sent a standardized set of questions about workforce planning efforts and challenges to each M&O contractor and NNSA site office, and analyzed their written responses. We also interviewed NNSA and M&O human capital officials at each site about site-specific workforce challenges and their efforts to address them. We reviewed two NNSA systems for managing human capital data; to assess the reliability of these systems, we interviewed knowledgeable NNSA officials to assess the reliability of these data and determined that they were sufficiently reliable for the purposes of this report. We conducted this performance audit from December 2010 through April 2012, in accordance with generally accepted government auditing standards. Those standards require that we plan and perform the audit to obtain sufficient, appropriate evidence to provide a reasonable basis for our findings and conclusions based on our audit objectives. We believe that the evidence obtained provides a reasonable basis for our findings and conclusions based on our audit objectives.

BACKGROUND

To ensure the safety, security, and reliability of the nation's nuclear weapons stockpile, NNSA relies on contractors who manage and operate government-owned laboratories, production plants, and a test site. NNSA's eight enterprise sites each perform a different function, all collectively working toward fulfilling NNSA's nuclear weapons-related mission. Figure 1 shows the locations of the sites and describes their functions.

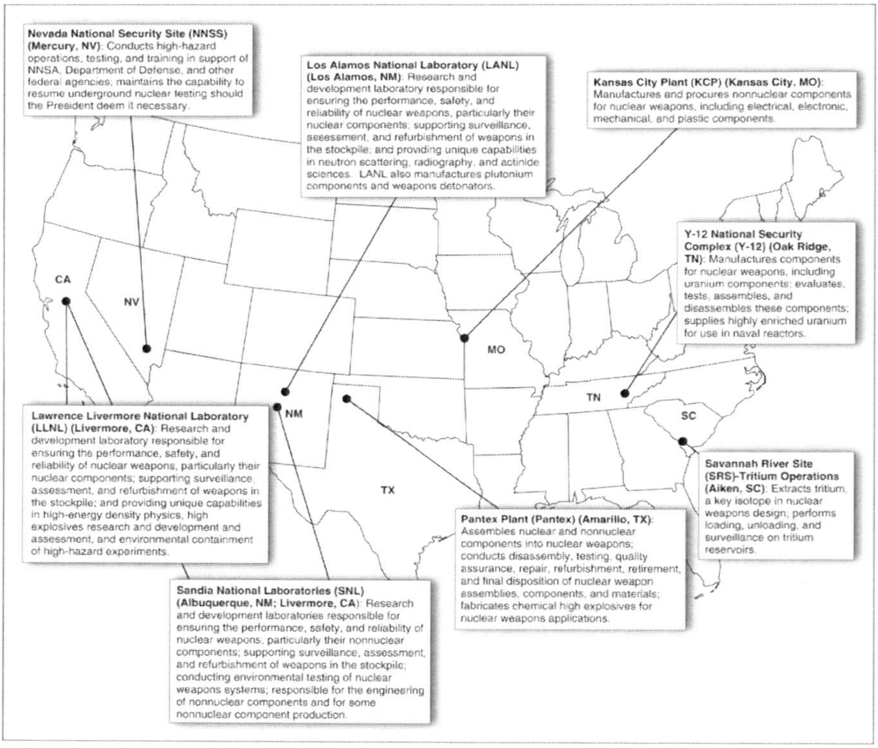

Source: NNSA; Map Resources (map).

Figure 1. Nuclear Security Enterprise Sites.

To provide support and oversight, NNSA locates between about 30 and 110 NNSA staff in a site office at each facility, and also draws on the resources of NNSA staff in headquarters and the Albuquerque complex. According to NNSA officials, this support and oversight requires that some NNSA staff have critical skills comparable to the contractors they support and oversee. For example, NNSA staff may need technical knowledge and expertise to accept and review deliverables from M&O contracts and, when presented with options, be able to determine how best to proceed to meet contract goals, mission, and objectives. They may also need skills related to the safe operation of sensitive defense nuclear facilities such as expertise in occupational safety and fire safety. For example, according to NNSA officials at the Livermore Site Office, most of the staff in critical skills positions there are focused on ensuring safety at the laboratory's nuclear facilities.

Maintaining critical skills within its workforce is not a challenge unique to NNSA. Every 2 years, we provide Congress with an update on GAO's high-risk program, under which GAO designates certain government operations as high risk due to their greater vulnerabilities to fraud, waste, abuse, and mismanagement, or their need for transformation to address economy, efficiency, or effectiveness challenges. In 2001, GAO designated strategic human capital management across the entire federal government as a high-risk area, in part because critical skill gaps could undermine agencies' abilities to accomplish their missions. We have also reported in the past that NNSA and its predecessor organizations' record of inadequate management and oversight of contractors has left the government vulnerable to fraud, waste, abuse, and mismanagement. Contract management at DOE has been on GAO's high risk list since 1990, the first year our high-risk list was published.[8] Progress has been made, but NNSA and DOE's Office of Environmental Management remain on our high-risk list.[9]

As of 2011, our most recent update of the high-risk list, significant steps had been taken to address some of the federal government's strategic human capital challenges. Strategic human capital management was designated a high-risk area 10 years earlier governmentwide and remains on the high-risk list because of a need for all federal agencies to address current and emerging critical skills gaps that are or could undermine agencies' abilities to meet their vital missions. Specifically, across the federal government, we reported that resolving remaining high-risk human capital challenges will require three categories of actions:

- *Planning.* Agencies' workforce plans must define the root causes of skills gaps, identify effective solutions to skills shortages, and provide the steps necessary to implement solutions.
- *Implementation.* Agencies' recruitment, hiring, and development strategies must be responsive to changing applicant and workforce needs and expectations and also show the capacity to define and implement corrective measures to narrow skill shortages.
- *Measurement and evaluation.* Agencies need to measure the effects of key initiatives to address critical skills gaps, evaluate the performance of those initiatives, and make appropriate adjustments.

NNSA and Its M&O Contractors Use Multifaceted Strategies to Recruit, Develop, and Retain Their Workforces

NNSA and its M&O contractors have developed and implemented multifaceted strategies to recruit, develop, and retain both the federal and contractor workforces needed to preserve critical capabilities in the enterprise. NNSA focuses on attracting early career hires with competitive pay and development opportunities, and the agency is reassessing future enterprisewide workforce needs. M&O contractors' strategies vary from site to site, but each site focuses on maintaining competitive compensation packages.

NNSA Focuses on Attracting Early Career Federal Workforce Hires with Competitive Pay, Flexible Schedules, and Development Opportunities

NNSA takes various steps to recruit, develop, and retain a federal workforce with the necessary critical skills. NNSA's recruitment strategies are focused primarily on students and recent graduates in science and engineering programs. NNSA generally relies on two key programs to develop its critically skilled workforce—one that identifies needs and another that identifies the qualifications necessary to meet them. Its retention efforts focus on competitive pay, flexible schedules, and development opportunities. NNSA is also undertaking a comprehensive reassessment to ascertain future federal workforce requirements.

NNSA's Recruitment Strategies Focus on Students and Recent Graduates in Science and Engineering Programs

NNSA has several programs targeted toward recruiting students and recent graduates, primarily in science and engineering fields. NNSA began these programs within the past 7 years as a means of succession planning.[10] NNSA's programs focused on recruiting students include the following:

- The Student Temporary Employment Program is a summer internship program for high school through graduate students of any discipline. Students participating in this program receive a salary while working at NNSA.
- The Student Career Experience Program is a program for graduate students in science, engineering, and other fields. This program aims to persuade skilled graduates to pursue careers in NNSA. Participants work for NNSA full-time during school breaks and part-time the rest of the year. These positions can be converted to full-time competitive appointments when participants receive their degrees.
- The Minority Serving Institutions Program aims to strengthen the diversity of the applicant pool by exposing younger minority students to technical fields and NNSA work early in their educational careers. This program focuses on students beginning in junior high school and continues through college entry and has cooperative agreements to enhance science, technology, engineering, and mathematics curricula at all levels at 29 minority-serving institutions. Since the program's inception in 2007, 167 minority students have participated in hands-on research at NNSA site offices and laboratories.

NNSA's key program for recruitment of recent graduates is its Future Leaders Program. NNSA established the program in 2005 to recruit recent U.S. citizen graduates of bachelor's and master's programs, primarily in engineering and science. The Future Leaders Program is a 2- year development program that requires participants to complete classroom and on-the-job training, in addition to developmental assignments outside their home office. NNSA hires about 30 recent graduates into this program each year. Applicants are hired into the program offices where they will be permanently placed and are selected based on each program office's skills needs. According to NNSA officials, approximately two-thirds of the 175 program participants hired from 2005 through 2010 have engineering and science backgrounds that enable them to develop the technical critical skills NNSA needs to provide support and oversight of contractors. As their careers advance, some program participants are expected to become more focused on developing deep expertise in a particular technical area, and others will gravitate toward more senior management and leadership positions. NNSA officials told us they consider the program very successful because nearly 90 percent of all those hired into the program since 2005 remained at NNSA.

NNSA's Development Strategies Include Two Key Programs

NNSA relies primarily on two programs to develop a federal workforce with the requisite critical skills—the Federal Technical Capability Program and the Technical Qualification Program (TQP). NNSA employees' critical skills generally fall into two broad categories: (1) technical skills related to managing the safe operation of nuclear facilities, and (2) technical knowledge and expertise necessary to accept and review contract deliverables. To ensure that it has sufficient numbers of federal employees with critical skills to manage the safe operation of nuclear facilities, NNSA relies on the Federal Technical Capability Program[11]—a DOE-wide effort to define requirements and responsibilities for meeting the department's commitment for recruiting, developing, and retaining the technically competent workforce necessary to achieve this mission. To implement the goals of the Federal Technical Capability Program at the site level, NNSA senior managers conduct annual workforce analyses and develop staffing plans that identify critical technical capabilities and positions that ensure the safe operation of nuclear facilities. For example, NNSA relies on senior managers to identify the fire safety needs for the National Ignition Facility, a stadium size research facility at Lawrence Livermore National Laboratory and to identify how many fire protection engineers are required to meet these needs. To help meet these goals, DOE established the TQP, which sets technical qualification requirements for NNSA positions related to the safe operation of nuclear facilities and tracks federal employees' progress in meeting these qualifications. More specifically, the TQP documents how NNSA:

- identifies needed positions,
- tailors qualification standards for them,
- establishes time and duty limitations for qualification,
- describes the process to identify learning activities to achieve competency for the specific job duties, and
- establishes methods for evaluating qualification.

NNSA officials told us that only federal employees in positions related to managing the safe operation of nuclear facilities are required to participate in the TQP. However, NNSA managers may also subject employees who accept and review contract deliverables to TQP requirements to help ensure that they have the skills necessary to evaluate technical criteria of contract deliverables.

Beyond the TQP, according to NNSA officials, human capital managers rely on annual human capital needs assessments to inform subsequent recruitment and hiring efforts to ensure the requisite mix of skills is present in the federal workforce. These assessments consider attrition and other demographic data, succession planning, and education and experience requirements. For example, NNSA officials told us that in 2011 its Office of Human Capital Management surveyed NNSA programs to identify needs for the Future Leaders Program. As part of this survey, they analyzed attrition in the federal workforce and used the information to assist in decisions about how many engineers to hire across the enterprise through the Future Leaders Program. Recruitment efforts in 2012 will focus on finding replacements for these engineers.

NNSA's Retention Strategies Focus on Competitive Pay, Flexible Schedules, and Career Development Opportunities

NNSA's retention strategies focus on offering new staff competitive pay, flexible schedules, and career development opportunities.

Competitive pay. According to NNSA officials, NNSA's retention efforts place a high priority on preserving the agency's capacity to offer competitive compensation. For example, for relatively new hires, such as those hired through the Future Leaders Program, NNSA can sometimes offer as much as $6,000 in lump sum hiring bonuses and up to $10,000 in student loan repayment in return for signing a service agreement. In some cases, NNSA is also able to offer retention bonuses of up to 25 percent of annual salary to employees that might otherwise leave federal service. In addition, NNSA has the flexibility to offer particularly desirable applicants higher starting salaries and reward top performers with higher pay.

For more senior employees, according to NNSA officials, DOE and NNSA sought, and were granted, authorities by Congress to offer higher pay to staff primarily in certain engineering and science fields. Specifically, to help it retain more experienced competitive service employees with critical skills— that is, employees in regular civil service positions— Congress granted exceptions to normal hiring regulations, including salary caps, under three excepted service authorities. First, under the Department of Energy Organization Act,[12] the Secretary of Energy is granted special excepted service hiring authorities to hire up to 200 highly skilled scientific, engineering, professional, and administrative individuals to upgrade the department's technical and professional capabilities. NNSA can use this authority in some cases to hire senior-level employees from outside the government or difficult-to-hire administrative staff. According to NNSA officials, there are presently 50 such individuals employed by NNSA. Second, under the National Defense Authorization Act,[13] the Secretary of Energy is also granted special excepted service hiring authorities to hire up to 200 highly skilled individuals—typically scientists, technicians and engineers with skills related to and necessary for the operation of nuclear facilities. According to NNSA officials, there are currently about 100 such individuals currently employed by NNSA. Third, under the National Nuclear Security Administration Act,[14] NNSA may hire up to 300 highly qualified scientists, engineers, and other technically skilled workers needed to support the missions of NNSA under similar excepted service hiring authorities. According to NNSA officials, NNSA has used this authority to hire and employ about 280 highly skilled individuals.[15] NNSA officials told us that all of these flexibilities are useful and help NNSA compete with the Nuclear Regulatory Commission and national laboratories.

Flexible schedules. NNSA's retention efforts also include a flexible schedule program that gives employees the opportunity to work a nontraditional schedule or vary their work hours from day to day. For example, employees with school-aged children may opt to work more than 8 hours some days and fewer hours other days in order to accommodate school early release days.

Development opportunities. NNSA offers some employees the opportunity to undertake career development opportunities such as rotational assignments and details. Integral parts of the Future Leaders Program are 30-day local rotational assignments and 60-day headquarters or field assignments away from their home locations. For example, a Future Leaders Program participant based in NNSA's Washington headquarters who is interested in a program run by Sandia National Laboratories in Albuquerque might be assigned for 60 days to related work at NNSA's Sandia Site Office or Albuquerque complex. In addition, NNSA has implemented a program called the In-Teach Program, which focuses on knowledge preservation and transfer by providing funding to train highly skilled senior employees to become more adept at transferring knowledge and skills to less skilled more junior employees.

NNSA Is Undertaking a Comprehensive Reassessment to Ascertain Future Federal Workforce Requirements

NNSA is currently undertaking a comprehensive reassessment and analysis of the staffing requirements for its federal workforce through 2016 in headquarters and field locations. NNSA officials told us that the reassessment is needed for strategic planning purposes and will improve NNSA's efforts to ensure that its federal workforce has the skills necessary to carry out its missions, including technical, support, and oversight capabilities. The reassessment includes the following four phases:

- Describing and identifying organizational core competencies, and the workforce required for NNSA's future
- Analyzing the current workforce and gaps related to requirements for NNSA's future
- Developing a plan to close gaps between future requirements and the current workforce
- Developing and implementing a workforce management system which is integrated with legacy Department of Energy human capital information technology systems

NNSA officials told us they expect the reassessment and resulting report to be complete in fiscal year 2013.

M&O Contractors' Recruitment, Development, and Retention Strategies Vary but Generally Focus on Competitive Compensation Packages

M&O contractors' recruitment, development, and retention strategies are site-specific. Generally, their recruitment efforts vary by the type of employee needed—particularly, whether the position requires an advanced degree. Their development efforts vary in approach but are also site specific and face some challenges—particularly in preserving underground nuclear testing skills. Their retention efforts focus on maintaining competitive total compensation packages—salaries and benefits—but their strategies to mitigate attrition vary from site to site.

M&O Contractors' Recruitment, Development, and Retention Strategies Are Site-Specific

NNSA's M&O contractors have developed and implemented site-specific strategies to recruit, develop, and retain the workforces needed to preserve critical capabilities throughout the enterprise and accomplish NNSA's mission. Accordingly, contractors have typically developed site- specific workforce planning systems that enable them to identify the kinds of candidates they need to recruit, develop, and retain in order to align projected nuclear weapons-related work and budget resources. Using these workforce planning systems, site managers can anticipate the nuclear weapons-related work NNSA has contracted for, how it will be funded, how many staff are required, and what skills will be needed, and can avoid potential shortages in staff or skills. For example, in the course of their 2- and 5-year planning processes, managers at Sandia National Laboratories use a four-step workforce planning tool, the Workforce Acquisition Project, to anticipate critical skills hiring needs based on the expected lab-wide business outlook and attrition. This early assessment of critical skills requirements ensures that the contractor has sufficient time to identify and recruit new staff as necessary and give new staff time—generally 2 to 5 or more years—to develop their skills.

M&O Contractors' Recruitment Strategies Vary According to the Types of Employees Needed

M&O contractors' strategies for recruitment vary according to the kinds of employees they need to hire—in particular, whether the position requires an advanced degree. For example, the weapons laboratories, which include Sandia, Los Alamos, and Lawrence Livermore National Laboratories, typically require highly skilled candidates with advanced degrees to replace physicists, engineers, and other experts who retire or leave for other jobs. M&O contractors at weapons laboratories thus focus their recruitment efforts on students and recent graduates of the nation's leading graduate schools in science, engineering, and mathematics. Efforts to attract candidates begin with summer internship programs and continue with support for post-doctoral fellowships and direct offers of employment. Officials at the Lawrence Livermore National Laboratory told us that, in addition to these efforts to recruit students and recent graduates, they also recruit at the midcareer or higher level at professional meetings in science and technology fields and through the cooperative relationships with American universities and industries to broaden the prospective employee pool and enhance the intellectual vitality of its existing workforce.

According to M&O contractor officials, the critical skills needs at other enterprise production plants, such as the Y-12 National Security Complex and Pantex Plant, differ from those at the weapons laboratories, and their recruiting strategies reflect these differences. Unlike the weapons laboratories, production plants generally do not require candidates with advanced degrees; rather candidates typically need a bachelor's degree or, in the case of manufacturing and skilled craft positions, an associate's degree or skills in advanced manufacturing techniques. As such, M&O contractors at production plants can generally recruit regionally for the staff they need and have less need to recruit nationally. For example, M&O contractor officials at Y-12 told us that they recruit predominately bachelor's level candidates—predominately engineers—from universities within a 300-mile radius of Oak Ridge, particularly from the University of Tennessee in nearby Knoxville. Production plants are also generally well- established within their communities and focus most of their recruitment efforts for skilled manufacturing positions on the local area. For example, M&O contractor officials at the Pantex Plant told us that they have developed strong ties with local community colleges over the years and typically look for high school graduates and community college students and graduates with some specialized, skilled training or work experience. Nevertheless, according to Pantex officials, they have also taken advantage of opportunities to recruit from outside the local areas, seizing opportunities to recruit automotive workers with machine tool experience and highly skilled plant workers from another nuclear security enterprise production facility, the Savannah River Site, in the wake a reduction in force.

M&O Contractors' Development Strategies Vary in Approach and Scope and Face Some Challenges

M&O contractors told us their strategies for development are often linked to recruitment because appealing development opportunities can encourage candidates to accept job offers. As with strategies for recruitment, those for development are tailored to the specific needs of each site's workforce, but many of the M&O contracting officials we spoke with cited continuing educational opportunities and the option to move within the organization as appealing development opportunities. For example, M&O contractor officials at Sandia National Laboratories told us that offering continuous training opportunities and the opportunity to move to different jobs within different components of the laboratory was very appealing to entry-level hires. Accordingly, Sandia's Corporate Learning and Professional Development Programs offer various training opportunities. Sandia officials told us that these opportunities help employees keep skills current, provide additional educational opportunities, and help laboratory management anticipate critical skills needs in the workforce. As part of these programs employees can also take training offered by Sandia's technical and compliance training group, which is focused on skills currently in demand at Sandia, or participate in university graduate degree programs, which Sandia will pay for. The Lawrence Livermore National Laboratory's Education Assistance Program provides up to $50,000 in tuition assistance for coursework toward a higher degree.

Production plants also offer continuous learning and development opportunities. For example, the M&O contractor officials at the Kansas City Plant told us employees are encouraged to pursue higher education in areas where the plant has a skills gap. In such cases, the contractor will pay tuition and, if the employee attends school full-time, continue to pay 70 percent of the employee's base salary. Kansas City Plant employees may also participate in developmental programs at the entry or midcareer levels that allow participants to undertake three rotational assignments to support their targeted and tailored personal development plans. In addition, the Pantex plant offers employees support for technical training opportunities with local colleges. The Nevada National Security Site also offers a number of developmental opportunities to its staff, including a voluntary mentoring program for all employees, assistance with career planning, various training and certification programs, and attendance at seminars and conferences. M&O contractor employees also have access to online courses and books as well as CD-based training sessions on a wide variety of topics, including supervision, management and leadership; computer skills and certifications; communication; and mentoring.

M&O contractors told us that, in their development efforts, they rely on knowledge preservation and transfer programs, including recording the performance of high-skill critical tasks, formal classroom training, on-the- job training, and mentoring programs to preserve critical capabilities in the nuclear security enterprise. Knowledge preservation programs are focused on the physical preservation or recording of critical information and knowledge—typically in paper records, microfilm and microfiche, and in various audio and video media. Knowledge transfer programs seek to ensure that experienced laboratory or production plant employees successfully pass on the knowledge to replicate critical tasks to newer employees.

Knowledge preservation. All M&O contractors at nuclear security enterprise sites have taken steps to record critical knowledge. These knowledge preservation programs are broadly similar from site to site, whether laboratory or production plant. For example, Los Alamos National Laboratory officials report that their archives house information on weapons designs and experiments dating to the inception of the laboratory. This information is contained in documents and other media such as film, audio and videotape, drawings, and photographs. The information housed in the archives is still relevant and is used by researchers across the enterprise. It may also be used outside the enterprise by, for example, documentary filmmakers and occupational health researchers. More recently, in the 2000s, Los Alamos gathered and developed critical information in the course of the Reliable Replacement Warhead Program—a program that explored the possibility of developing new nuclear weapons designs. Los Alamos engineers and scientists documented all decisions in the Reliable Replacement Warhead design process through written and video documentation. The other weapons laboratories have also invested in electronic records and videos to preserve critical knowledge. According to Lawrence Livermore National Laboratory officials, Livermore maintains an extensive electronic archive of papers and reports, as well as tutorial lectures by experienced weaponeers on key areas of weapons knowledge. Sandia National Laboratories also has its Knowledge Management Streaming Assets Library program, which has recorded about 1,500 hours of classified exit interviews with retiring weaponeers and made them available to current staff.

M&O contractors at the weapons production plants report broadly similar efforts to preserve critical knowledge at their sites. For example, the Y-12 National Security Complex has the Knowledge Preservation Program (KPP). Similar to Sandia National Laboratories' knowledge preservation efforts, the KPP films retiring employees as they do their work and interviews them on how they do it, then archives the videos in an electronically searchable format. As employees approach retirement, a KPP video and interview is part of the retirement checklist. These videos are evaluated for accuracy by an expert before they are entered into the KPP system. Y-12 officials told us that other NNSA sites have created videos or archives for knowledge preservation but they are not as easily accessible. M&O contractors at the Pantex Plant have undertaken similar efforts, including creating and maintaining what Pantex officials call "picture books" on weapons assembly, and interviewing experienced Pantex workers to capture their knowledge in areas such as high explosives and making these interviews available as a training tool.

According to M&O contractor officials at the Nevada National Security Site, however, efforts to preserve critical knowledge regarding underground nuclear testing have faced challenges, as they have been limited and sporadic. These efforts have been complicated by two factors: (1) the need to protect vital national security information against unauthorized disclosure led to a practice of not keeping written documentation about the specifics of critical tasks; and (2) significant numbers of employees were laid off in the mid-1990s after U.S. underground nuclear testing ended. Until 2007, NNSA maintained a program that undertook substantial efforts to capture and record critical knowledge possessed by these workers, but NNSA and M&O contractor officials said these efforts were not comprehensive or systematic, and funding was discontinued.

Knowledge transfer. M&O contractors at the weapons laboratories rely on a range of approaches to transfer knowledge, while there is more similarity among the knowledge transfer programs of M&O contractors at production plants. Specifically, each of the three weapons laboratories uses a combination of classroom training, on-the-job training, and mentoring relationships to transfer critical nuclear weapons design information, but with varying reliance on each of these three components. For example, at one end of the spectrum, Sandia National Laboratories relies most heavily on a classroom-focused curriculum—its highly regarded Weapons Intern Program. According to Sandia officials, the 11- month Weapons Intern Program succeeds in transferring such knowledge and experience through a blended learning environment, consisting of live and multimedia-based classroom instruction, individual and team research projects, hands-on activities, and off-site facility and operations tours and briefings. The live instruction is provided through a large contingent of subject matter experts in the various weapon technology, design, evaluation, production, operations, policy, and management areas.

Lawrence Livermore National Laboratory is at the other end of the spectrum, relying mostly on mentoring programs and on-the-job-training opportunities to transfer advanced nuclear weapon design skills to new staff. According to Lawrence Livermore officials, their approach to developing critical skills expertise is to embed new employees into work groups directly engaged in important work, with an experienced employee acting as a mentor. As new employees gain skills and experience and demonstrate their readiness, they are assigned tasks of increasing levels of complexity and responsibility. Laboratory officials stated that, in their experience, employees supporting the weapons program must be exposed to years of work in the field to acquire the needed knowledge and judgment to be a fully qualified weaponeer. An extensive electronic archive of papers and reports is available, as well as tutorial lectures on key areas of weapons knowledge, but Livermore officials told us there is no substitute for hands-on experience with weapons.

Los Alamos National Laboratory's approach is not as classroom-focused as Sandia's program, nor is it as dependent on mentoring relationships and on-the-job training as Lawrence Livermore's. Specifically, Los Alamos officials told us that critical skills are being transferred through a combination of formal training opportunities, mentoring, and archiving programs. For example, the TITANS program, referred to informally as "nuclear design university" is a 3-year, credential-granting program with 2 years of coursework and 1 year of thesis research and writing under the direction of a mentor. Thesis projects can either be focused on learning new modeling techniques or on mastering the simulation of above-ground experiments. For example, one knowledge transfer technique is to reanalyze old data from actual experiments to teach newer employees to use modern simulation techniques to estimate the results of real testing. The results of the student's analysis are then compared to actual testing data. Los Alamos officials told us this practice is a very effective method for examining how well the student has mastered the use of computer simulation techniques—a very critical skill when live nuclear testing is not an option.

Knowledge transfer at weapons production facilities is focused more on having employees demonstrate that they can replicate specific tasks. For example, M&O contractor officials at the Pantex Plant told us that they are very aggressively taking steps to ensure that younger workers can carry on performing some of the same tasks after older workers retire. The centerpiece of the Pantex effort is the Retiree Corps. Through this program, recent retirees are brought back on a part-time basis—for a maximum of 800 hours a year, an average of a little less than 2 days a week—specifically to teach current Pantex employees how to do their high-skill critical task. Retirees host talks and seminars, provide a narrative to schematics of detailed procedures and photos, and are recorded and/or videotaped explaining their tasks. Pantex officials told us they verify the knowledge transfer by requiring the trainee to demonstrate that he or she can replicate the task.

Again, however, the M&O contractor at the Nevada National Security Site faces some challenges. The site has an active on-the-job training program and specialized training on specific diagnostic and recording techniques relevant to underground nuclear testing. However, according to M&O contractors, funding for this program has been minimal for several years. In addition, according to Nevada National Security Site M&O contractor officials, it is challenging to preserve some of the critical skills necessary for underground nuclear testing when there is no opportunity to provide any direct experience with such testing.

M&O Contractors' Retention Strategies Focus on Maintaining Competitive Total Compensation Packages

NNSA officials and M&O contractors told us that maintaining competitive total compensation packages—that is, combined salary and benefits—is crucial for achieving their strategies for recruiting, developing, and retaining the workforce with the skills necessary to sustain critical capabilities in the nuclear security enterprise, but that other factors are also useful in both attracting desirable candidates and mitigating attrition. For example, M&O contractor officials at Sandia National Laboratories told us that offering the highest salary is not required to attract top talent, but offering pay comparable to peer institutions is a necessity. Accordingly, NNSA officials work very closely with M&O contractors to ensure that contractor compensation remains comparable to other enterprise laboratories and plants, private laboratories, companies, and other government entities that recruit and try to retain similar talent. M&O contractors undertake compensation studies every year and comprehensive benefits evaluation surveys every 2 years. This compensation study is done using survey data from recognized regional, national, and international surveys as needed. Based on these data, M&O contractors may seek permission from NNSA to pay certain employees more by submitting a special request in the Compensation Increase Plan. If the plan is accepted by NNSA, salaries will be increased. In addition to raising salaries for M&O contractors to keep them competitive, NNSA will also authorize and pay for sign-on and retention bonuses, significant monetary recognition and awards programs, and special compensation packages for especially difficult-to-recruit and retain critical skills specialties. The biennial benefits evaluation compares the value of M&O contractor workforce benefits to 15 peer competitors for the same talent.

According to DOE policy,[16] M&O contractors may offer benefits up to 105 percent of the value of peer institutions' benefits.

NNSA officials and M&O contractors told us that other factors are useful in both attracting desirable candidates and mitigating attrition. For example, the weapons laboratories in particular can offer scientists and engineers access to state-of-the-art equipment—such as the National Ignition Facility at Lawrence Livermore National Laboratory—and the opportunity to do cutting edge research that cannot be done outside the enterprise due to national security restrictions. Similarly, for the three production plants located in relatively remote, nonmetropolitan locations—particularly Pantex, Y-12, and the Savannah River Site— attrition rates are lower among candidates with ties to the local area. For example, M&O contractor officials at Y-12 told us that they recruit locally to the extent possible, because, historically, employees from nearby communities have been less likely to seek opportunities that would require them to relocate. These officials added that the local community is familiar with Y-12, and that about 35 percent of new applicants are employee referrals.

M&O contractors have broadly similar retention initiatives. While M&O officials at all sites in the enterprise told us that competitive total compensation packages—that is, salary and benefits—are ultimately the most important factors in employee retention, sites also typically offer a similar mix of other programs designed to encourage retention, such as work/life balance programs, flexible work schedules, and some form of continuous education and learning programs. In addition, some of the M&O contractors we spoke with told us that, to the extent they are able, they try to accommodate the desires and expectations of more recently hired employees for opportunities for faster advancement, meaningful and challenging assignments, and recognition of high performance.

NNSA Monitors Key Human Capital Metrics to Assess the Effectiveness of Strategies to Maintain Critically Skilled Workforces

To assess the effectiveness of its strategies for recruiting, developing, and retaining the NNSA staff and M&O contractors needed to preserve critical skills in the nuclear security enterprise, NNSA monitors key human capital metrics, including the length of time to hire employees and attrition. To assess the effectiveness of its M&O contractors' strategies, NNSA uses M&O

contractors' data to monitor key human capital metrics, but these metrics do not have standardized definitions.

NNSA Monitors Key Human Capital Metrics for Its Staff

To assess the effectiveness of its strategies for recruiting, developing, and retaining the federal workforce with the requisite critical skills to support and oversee M&O contractors, NNSA focuses on monitoring two key metrics—the length of time it takes them to hire an employee and its attrition rates—and tracks employees' progress toward completing the required training and certifications through the TQP. NNSA officials told us the length of time it takes them to hire an employee is a useful metric because it is an indicator of the efficiency of their human capital management processes. Attrition rates, especially for those leaving NNSA for reasons other than retirement are a valid indicator of the relative attractiveness of NNSA employment. Increases in the time it takes to hire employees and increases in the attrition rate would indicate a potential problem that would eventually make it more difficult for NNSA to attract and retain the workforce it needs to achieve its mission.

Overall responsibility for maintaining a federal workforce with the necessary critical skills to carry out NNSA's mission resides in NNSA's Office of Human Capital Management, located at NNSA headquarters, and its site offices are also responsible for closely monitoring changes in their workforces and keeping NNSA headquarters informed of any changes. They also have direct responsibility for making sure that site office employees are maintaining the technical certifications required to perform their duties. NNSA's Office of Human Capital Management Services, located at the Albuquerque complex, may also assist both headquarters and site office staff in monitoring these issues.

NNSA Monitors M&O Contractors' Human Capital Data, but Some Metrics Do Not Have Standardized Definitions

To assess the effectiveness of its M&O contractors' strategies for recruiting, developing, and retaining their workforces, NNSA monitors key human capital metrics using data the contractors collect. M&O contractors assess key human capital metrics, but these metrics do not have standardized definitions.

M&O Contractors Are Responsible for Managing Their Workforce Needs but NNSA Monitors Key Human Capital Metrics

NNSA generally gives M&O contractors the primary responsibility for identifying their workforce needs and taking the necessary steps to ensure they maintain workforces with the skills to meet the responsibilities outlined in their M&O contracts with NNSA. Accordingly, NNSA officials told us that, in 2005, they discontinued a requirement for M&O contractors to report on efforts to recruit and retain staff with critical skills, as well as more formal reporting requirements for workforce and succession planning. More specifically, according to NNSA officials, M&O contractors expect NNSA to instruct them on what they are required to do and what the contract deliverable and timeline is, but expect to be able to determine on their own how to meet their contractual obligations, including how to recruit, develop, and retain staff with the requisite critical skills.

Nonetheless, M&O contractors collect data on key human capital metrics for their workforces and provide these data to NNSA directly from their own human resource data systems. All contractors also undertake some level of workforce and succession planning, although there are no formal or specific requirements directing how they do so. According to NNSA officials, these metrics vary from site to site, but generally provide the same key information, including

- acceptance rates for offers of employment, which are benchmarked on a site-specific basis but are typically around 80 percent;
- attrition rates, both for retirement and non-retirement reasons, which are also benchmarked on a site-specific basis;
- pay comparability—whether salaries are competitive with peer institutions;
- benefits comparability—whether benefits are competitive with peer institutions; and
- ability to fill a critical skills position within a certain number of days—usually 48 to 90 days.

According to NNSA officials, these five metrics are tracked very closely by M&O contractors at all sites, and attrition, employment acceptance rates, and pay and benefits comparability data are systematically collected at regular intervals enterprisewide. If any of these metrics indicate a problem in retention, for example, NNSA officials told us, action would be taken to address it. For example, these metrics were monitored very closely by NNSA and the M&O contractors at Los Alamos National Laboratory and Lawrence Livermore National Laboratory during their 2006 transition to a new M&O contract with less generous retirement and medical benefits. There were concerns that this change could lead to a spike in attrition among highly skilled staff that could in turn lead to difficulties in the laboratories meeting deadlines on project deliverables. Similarly, NNSA is now carefully watching the same metrics at Sandia National Laboratories because the M&O contractor substantially cut future retirement benefits that took effect for those employees who remained at the lab beyond the end of 2011. If the metrics indicate greater attrition than expected, the laboratory could adjust its recruiting strategies to hire more staff.

NNSA also maintains close, cooperative working relationships between its federal and contractor workforces. Much of NNSA's expertise in M&O contractor human capital issues resides in its Contractor Human Resources Division (CHRD) at its Albuquerque complex. According to NNSA officials, the work of CHRD is both critical and central to how NNSA manages human capital issues with the M&O contractors. CHRD staff are in day-to-day contact with the M&O contractors on a wide range of human capital issues, including those related to recruitment, development, and retention of employees with critical skills. For example, if an M&O contractor is having difficulty recruiting staff with particular critical skills, it can submit a supplementary Compensation Increase Plan to the NNSA site office for authorization to offer candidates higher salaries. When this occurs, NNSA headquarters and the relevant site office largely rely on CHRD to review, analyze, and make recommendations to senior management on whether to accept, amend, or reject such a request. Because most sites do not have full-time human capital subject matter expertise in residence, NNSA site office officials in particular rely heavily on CHRD both for such expertise and to monitor M&O contractors' human capital performance metrics at all nuclear security enterprise sites. For example, officials at the Sandia Site Office told us that there is no full-time subject matter expert on human capital issues at the site office, so the office relies heavily on a CHRD staff member to inform the office's oversight of Sandia National Laboratories on this issue. According to NNSA officials, if NNSA had concerns about what a contractor was doing or had doubts that the contractor was going to be able to continue meeting its contractual obligations because of weaknesses in its recruitment, development, and retention strategies for critically skilled workers, NNSA would raise such concerns and require that corrective actions be undertaken.

However, as we noted in our February 2011 report,[17] NNSA lacks comprehensive information on the status of its M&O contractor workforce. Specifically, the agency does not have an enterprisewide workforce baseline of critical human capital skills and levels for the M&O contractor workforce to effectively maintain the capabilities needed to achieve its mission. NNSA officials said this is primarily because NNSA relies on its contractors to track these critical skills. As a result, we recommended that NNSA establish a plan with time frames and milestones for the development of a comprehensive contractor workforce baseline that includes the identification of critical human capital skills, competencies, and levels needed to maintain the nation's nuclear weapons strategy. NNSA stated that it understood all of our recommendations in that report and believed that it could implement them. NNSA has taken some actions toward this recommendation. As of March 2012, NNSA had completed a draft plan and was incorporating stakeholders' comments. NNSA officials said that they expect to complete the final contractor workforce baseline plan by May 2012. While contractor efforts may be effective at a specific site, these efforts neither ensure long-term survival of these skills across the enterprise nor provide NNSA with the information needed to make enterprisewide decisions that have implications on human capital. NNSA officials told us that they have determined that, as the responsible federal oversight agency for its M&O contractors, they recognize that they need a comprehensive and enterprisewide outlook regarding M&O contractor workforce data, particularly the identification of the critical skills needed to maintain and sustain future capabilities, and to verify that strategies are, indeed, in place to meet future requirements.

Accordingly, NNSA officials told us that they are developing the Enterprise Modeling Consortium—an initiative to, among other things, develop the needed skills data and models necessary to help NNSA manage its contractor workforces in a more proactive manner. The consortium is designed to help NNSA undertake more integrated, enterprisewide M&O contractor workforce reporting and analysis and identify the skills and competencies needed by the workforce, as well as the necessary staffing levels, based on the known and projected integrated program requirements needed to implement the Stockpile Stewardship Management Plan and associated budgeted programs for NNSA, DOE, and other federal agencies. NNSA officials told us that NNSA provided $400,000 to the Enterprise Modeling Consortium in fiscal year 2012 to fund further research and development on modeling. However, according to these officials, there is significant work left to do on the Consortium and they cannot provide an estimate for when the Consortium will be completed.

M&O Contractors Assess Key Human Capital Performance Measures, but These Metrics Do Not Have Standardized Definitions

Each M&O contractor collects key human capital performance data; however, we found that there are no specific, enterprisewide definitions of these data. NNSA officials told us that they have not asked M&O contractors to standardize these definitions because they believe their current system is effective. We previously reported that the lack of standard definitions for performance measurement data can significantly hinder agencies' ability to use such data in planning and reporting.[18] NNSA officials also told us that they believe M&O contractors have effectively used the flexibilities provided in their contracts and have demonstrated that they can identify specific critical skills needed and take the steps needed to, by and large, sustain them. However, NNSA is now considering developing a more comprehensive enterprisewide system, the Enterprise Modeling Consortium, to track M&O contractor human capital performance metrics and other workforce data and common definitions of performance metrics may become more important. Specifically, without common enterprisewide definitions of human capital performance metrics, NNSA may not be able to collect consistent and comparable data across all eight sites in the enterprise. For example, one of the M&O contractors' key metrics—acceptance rates for offers of employment—may not be consistently measured across the enterprise. Human capital staff at one national laboratory told us they participated in a program they compared to "speed dating," whereby candidates at a career fair may be interviewed for multiple positions and given offers of employment on the spot. However, job applicants may receive multiple offers of employment in a single day and may accept more than one offer to negotiate for a better salary or to have more time to consider their options. In such a situation, the employment offer to a candidate could be counted as an acceptance even if that candidate never became a laboratory employee. When asked about this scenario, NNSA officials stated that it was their understanding that M&O contractors were only counting as accepted offers those who ultimately reported for work, but acknowledged there was no NNSA standard definition and that they did not know for certain how such offers were counted.

Successful human capital management and workforce planning depend on valid and reliable data. These data can help an agency determine: performance objectives, goals, and the appropriate number of employees, and can help develop strategies to address gaps in the number, deployment, and alignment of employees. However, NNSA has not identified or considered the potential inconsistencies in these human capital metrics; therefore, decision makers are relying on information that may not be consistently reported.

NNSA and Its M&O Contractors Face Challenges in Recruiting, Retaining, and Developing Their Workforces

NNSA and its M&O contractors face challenges in recruiting, retaining, and developing their workforces and are using several tools to address these challenges. NNSA and its M&O contractor work environments, site locations, and high costs of living pose recruiting challenges. NNSA and its M&O contractors also face shortages of qualified candidates, an aging workforce, and variable funding. NNSA and its M&O contractors are taking actions to address their current human capital challenges, where possible.

NNSA and Its M&O Contractor Work Environments, Site Locations, and High Costs of Living Pose Recruiting Challenges

Officials from NNSA site offices and M&O contractor work sites reported that their secure work environment and location make recruitment of advanced science and technology candidates more challenging. Due to the sensitive nature of nuclear weapons work, NNSA and M&O contractor sites must be more secure than most private sector laboratories or commercial plants. To meet this security requirement, laboratories and plants in the enterprise tend to be restrictive environments, isolated from security threats by geography and classification protocols. In addition to these potentially undesirable traits, in the view of some candidates, some sites are further constrained by a high cost of living.

Restrictive environment. Officials from most M&O contractors reported that the restrictive environment required for nuclear weapons research and maintenance is a disadvantage in recruiting new staff with the potential to become weapons experts. Staff typically need to acquire and maintain high-level clearances and must often work in secure areas that prohibit the use of personal cell phones, personal e-mail, and social media. In particular, they told us younger candidates typically expect to stay continuously connected to their peers via cell phone and social media. Furthermore, any research completed in classified work can only be seen within the classified community; for researchers who desire broader recognition of their work and opportunities for wider collaboration, academia or private industry may be more attractive. Because of these restrictions, most M&O contractor human resources staff told us that it was more difficult to recruit younger scientists and engineers.

Isolation. An isolated location may be desirable for building or maintaining nuclear weapons, but it may not appeal to some desirable candidates with advanced degrees in science, technology, and engineering. For example, Los Alamos National Laboratory officials told us that the laboratory's relative isolation—nearly 100 miles from Albuquerque, New Mexico— may make it less appealing to some candidates. In addition, the relative lack of other types of employment opportunities nearby may pose challenges for candidates with spouses in careers outside of science, technology and engineering. Officials at two of the three weapons laboratories told us they focus on recruiting top candidates nationwide to gain a wide breadth of thought and opinion among their staff. The laboratories track the proportion of job offers accepted but cannot always ascertain or be sure of the reason a candidate rejects an offer because, according to officials at Lawrence Livermore, candidates may simply state they declined an offer for "personal reasons."

In addition, some of the production plants and the test site are also in isolated locations and face some of the same challenges as the laboratories. However, these sites require fewer candidates with advanced degrees and can generally rely on the local workforce to fill other types of critical skills positions. For example, Savannah River Site and Pantex are also both located far from other large cities. However, because of their relative isolation, they are among the biggest employers in these areas, and many local candidates are qualified and eager to accept positions in weapons manufacturing and maintenance. Pantex officials reported that they do not have difficulty finding most workers to perform weapons maintenance, which requires a shorter amount of on- the-job training than weapons design but nonetheless requires a set of critical skills. However, site staff have had to develop strategies to attract candidates to fill those positions that require advanced degrees. Unlike the laboratories, officials at all of the production plants told us that they focus their recruiting efforts for these positions at local and regional colleges and universities. Officials at Y-12, for example, have identified competitive science and engineering programs at universities within 300 miles of their plant in Oak Ridge, Tennessee. Y-12 officials reported that they have better results in both recruiting and retaining critically skilled workers when those workers have personal ties to the area. In contrast, M&O contractor officials from the laboratories told us that they needed to recruit from the top academic programs across the country.

High cost and competition. Two enterprise sites are located in areas with high costs of living, which can deter qualified candidates—Los Alamos and Lawrence Livermore. NNSA and its M&O contractors have flexibility to offer higher compensation for some critical skills, but some candidates are unwilling to live in high cost areas. For example, housing in Los Alamos is expensive and scarce. According to Los Alamos National Laboratory staff, some employees commute nearly 100 miles each way from Albuquerque every day partly due to cost of living constraints. Los Alamos Human Resources managers reported that high housing costs are a concern among current and prospective employees. Lawrence Livermore National Laboratory, located in the San Francisco Bay Area, is also a high cost area. NNSA has authorized higher salaries for some critically skilled M&O contractor employees but delays during the hiring process can give private sector recruiters an advantage with critically skilled candidates. Lawrence Livermore uses the flexibilities it has to negotiate competitive compensation, but a candidate interested in weapons work may be drawn to another site with a lower cost of living, such as Sandia National Laboratories in Albuquerque or one of the production plants.

Further complicating NNSA's recruiting efforts is the demand for qualified candidates in the private sector as well, and private sector jobs may offer a work environment that many candidates may find more desirable. The same pool of candidates who can excel in engineering, modeling, and simulation tasks is also attractive to high technology firms. For example, according to M&O contractor officials at Lawrence Livermore National Laboratory, a web-based provider of DVD rentals and streaming media uses computational scientists to predict consumers' preferences for films, which is the same skill set the weapons laboratories would use for modeling and simulation. However, this company does not have the constraints that a federal contractor has with compensation limits and a restrictive work environment.

NNSA and Its M&O Contractors Face Shortages of Qualified Candidates, an Aging Workforce, and Variable Funding

NNSA and its M&O contractors are making workforce plans, but face shortages in qualified critically skilled candidates and an aging workforce. In addition, uncertainty about future funding makes long-term workforce development initiatives challenging to execute.

Qualified Candidates Are in Short Supply and the Current Workforce Is Aging

The laboratories have not yet experienced any critical shortages of critically skilled workers, but they all reported that finding candidates with the appropriate qualifications is a growing recruitment challenge and that a more mobile and aging workforce is a retention challenge.

Shortages of qualified candidates. NNSA officials told us that qualified candidates are in short supply and that competition from science and technology-related companies in the private sector poses additional challenges. Candidates for most critical skills positions at national laboratories must meet certain criteria, including (1) an advanced degree (master's or doctorate) in a scientific, technical, or engineering field; (2) the ability to obtain a high-level security clearance, which requires U.S. citizenship; and (3) an interest in and willingness to learn weapons design work. The requirement for U.S. citizenship in particular is becoming an increasingly difficult criterion to satisfy in the recruitment process. National laboratory officials told us that a large percentage of students graduating from top science, technology, and engineering programs are foreign nationals. M&O contractors can hire foreign nationals to work outside of weapons-related areas, but the citizenship requirement for working on programs supporting U.S. nuclear weapons is not negotiable.[19]

In addition, national laboratory recruiting staff noted hurdles finding candidates with an interest in and willingness to learn weapons design work. For example, officials at Sandia National Laboratories told us younger candidates with the necessary qualifications are often more interested in fields that contribute to improving the environment. In addition, because of the sensitive nature of weapons work, civilian graduate programs cannot teach weapons-specific skills, so would-be weaponeers may not know whether the work suits them until after they have invested significant time working in the enterprise. Even if candidates accept a position, they do not actually have the authorization to design nuclear weapons; current policy allows them to refurbish components within the existing stockpile, and then only when funding is appropriated for that specific activity.

A more mobile workforce. NNSA and M&O contractor officials noted that a general shift from defined benefit retirement systems offering pensions to a defined contribution retirement system has made employees much more mobile and, therefore, harder to retain. A defined contribution retirement system makes employees much more mobile because, once the employee is vested—typically after a few years—their contributions to their retirement accounts are portable, therefore they no longer depend on tenure with a single employer. According to NNSA officials, M&O contractors no longer expect newly hired employees to spend their entire careers in the enterprise; rather, they expect them to work for a national laboratory or production plant for an average of 5 to 10 years.

Aging workforce. Many of the critically skilled employees currently filling these positions, both at the national laboratories and other NNSA sites, are at or near retirement age, which adds additional uncertainty to the projected human capital needs of the enterprise. NNSA officials told us that they are aware that many critically skilled employees are at or near retirement age, and they are tracking those retirements closely. Human capital staff from NNSA and its M&O contractors told us that it is difficult to anticipate retirement trends, especially during an economic recession. M&O contractor human resources staff said that they have found fewer staff retiring than they would have projected, due to uncertainties about their financial investments. These economic factors may have helped to preserve some critical skills within the enterprise, but officials are concerned that when the economy rebounds, eligible staff may retire at higher-than-projected levels. Such levels of attrition could leave a skills gap that would take years to replenish.

Variable Funding Can Impede Workforce Development Efforts

Knowledge transfer activities in the nuclear security enterprise tend to require multiple years to complete, but contractors have been challenged to plan and maintain these development efforts because funding varies from year to year. NNSA officials typically do not dictate whether or how much funding goes toward knowledge transfer within contractor workforces, except for specific programs at Sandia, because NNSA prefers not to fence funding for particular contractor activities. Contractors use what NNSA calls science campaigns—which, among other things, fund research to improve the ability to assess warhead performance without nuclear testing and help to maintain the scientific infrastructure of the nuclear weapons laboratories—and life extension programs—which ensure weapons' readiness and extend the life of existing warheads through design, certification, manufacture, and replacement of components—as a means for knowledge transfer, where more experienced weaponeers can train newer staff on weapons design and maintenance. Both science campaigns and life extension programs require long-term planning to ensure that the necessary resources are available.

According to NNSA and M&O contractor officials, funding for science campaigns and life extension programs has varied over the years. M&O contractor officials at both plants and laboratories told us their knowledge transfer plans have been adversely affected in years when funding has been reduced. In recent years, plans for certain life extension programs and science campaigns have been scaled back after plans have been made and contractor resources allocated. According to M&O contractors at the laboratories, reduced funding for life extension programs diminishes their opportunities to give their newer weaponeers hands-on experience. For example, weapons staff at Lawrence Livermore National Laboratory told us that they made knowledge transfer plans based on their approved warhead life extension projects, and that when those projects were sidelined; newer weaponeers were denied significant training opportunities. However, because funding decisions are beyond the M&O contractors' purview, M&O contractor officials told us there is little they can do to prepare for or mitigate this challenge.

NNSA and Its M&O Contractors Are Addressing Their Current Human Capital Challenges, Where Possible

NNSA and its M&O contractors reported that they are taking actions to address their human capital challenges where possible. Specifically, NNSA and M&O contractor officials told us they engaged in workforce planning to avoid potential critical skill gaps in the enterprise. NNSA-wide workforce plans are not expected to be completed until 2013 according to NNSA officials, but certain components are already in practice at various sites, such as streamlined hiring and security clearance practices and "pipeline" building for critically skilled employees.

Streamlined hiring and security clearance processes. NNSA and its M&O contractors have streamlined human capital processes to attract and hire new critically skilled workers. In the past, federal hiring processes have caused longer waits, both for candidates awaiting a decision and for human capital officials awaiting security clearances for new hires. M&O contractor staff reported that delays had previously allowed strong candidates to find other opportunities, or if candidates were hired and waiting for a clearance, they could lose interest in the position before they started. M&O contractor staff told us that finding work for hired-but- uncleared staff to complete was frustrating for both the new staff and their supervisors. NNSA has made reducing cycle time a priority, and officials from several sites reported that they have been able to hire and obtain clearances for employees more quickly in recent years.

Building a pipeline of critically skilled employees. Both NNSA and its M&O contractor officials acknowledge that, due to the long period required for developing some critical skills employees, they need to anticipate their critical skills needs for multiple years in the future. All sites have recruiting and development plans to preserve critical skills in their workforce, which they refer to as a pipeline. Sites use pipelines in two ways to avoid critical skills gaps. First, they use training and project assignments to ensure that critical skills are being developed and preserved in newer employees. For example, Lawrence Livermore has assessed its employees' skill sets and experience, so it knows which are currently performing essential operations more than 25 percent of the time—called core employees— and which are being prepared to perform those operations—called pipe employees. They can augment a pipe employee's expertise in an area if management sees a shortage of core employees in that skill set. Second, in recruiting activities, human resources staff may maintain information about potential future candidates for weapons programs, either with contacts made in internship, fellowship, and coop programs or by keeping records of interested candidates who were not hired. For example, Sandia is building a database of potential candidates, so that in the future it is not relying exclusively on that year's graduating class from the top science and engineering programs.

Succession planning can also inform pipeline decisions. M&O contractor officials at some sites said that they have begun to analyze potential skills gaps if a specific retirement or separation were to occur. Those M&O contractors who are undertaking these analyses can rely on managers' assessments of their employees or software packages designed to facilitate succession planning. M&O contractors told us that this kind of planning is currently used in management or leadership capacities, but in the future it could be applied to other areas such as critical skills capacities. Each M&O contractor has a unique way of implementing its pipeline, but M&O contractor officials from all sites told us they all realize the need to consider future retirements and mission requirements in their current hiring and development plans. For example, a senior M&O contractor manager at Sandia National Laboratories responsible for building the laboratories' talent pipeline told us that Sandia is facing unprecedented hiring needs due in part to expected increases in retirements. He expects to experience 33 to 50 percent attrition in the next 4 to 5 years, while the total number of Sandia employees will need to remain about the same. Accordingly, Sandia officials told us they expect to have hired approximately 3,100 new employees in the 3 years ending in 2012—about 800 in 2010, 1,100 in 2011, and 1,200 in 2012.

Some of the human capital challenges facing the enterprise are beyond the control of NNSA and its M&O contractors, and in these cases, NNSA has authorized increased compensation to help the sites acquire or retain the personnel they require. The site locations are fixed, and site staff cannot change the number of U.S. citizens completing graduate science and technology programs. Similarly, NNSA and its contractors have no choice but to adapt to the increased mobility of their staff resulting from the shift to a defined contribution retirement systems. To mitigate these challenges, NNSA and its contractors continue to offer financial incentives to recruit and retain critically skilled employees, with competitive starting salaries. The scale of these financial incentives can vary by location and position, but NNSA reported that this strategy has thus far been adequate for recruiting and retaining the talent they need.

CONCLUSION

NNSA and its M&O contractors have taken a number of useful steps to sustain critical skills in the enterprise in the face of several challenges. NNSA has begun to implement the recommendation we made in our February 2011 report to establish a plan with time frames and milestones for the development of a comprehensive contractor workforce baseline that includes the identification of critical human capital skills, competencies, and levels needed to maintain the nation's nuclear weapons strategy. However, while contractor efforts may be effective at a specific site, they do not provide NNSA with the information needed to make enterprisewide decisions that have implications on human capital. Without this information, NNSA's ability to monitor the effectiveness of its and its M&O contractors' strategies to recruit, develop, and retain the workforces needed to preserve critical skills may be hindered. In particular, without common enterprisewide definitions of human capital performance metrics, NNSA may not be able to collect consistent and comparable M&O contractor human capital data across all eight sites in the enterprise. Since NNSA is now considering developing a more comprehensive enterprisewide system to track data on critical skills through its Enterprise Modeling Consortium, this may be an opportune time to explore establishing common, uniform definitions for the human capital metrics used in this system.

RECOMMENDATION FOR EXECUTIVE ACTION

To improve NNSA's ability to monitor the effectiveness of its strategies—and its M&O contractors' strategies—to recruit, develop, and retain the workforces needed to preserve critical skills in the enterprise, we recommend that the Administrator of NNSA take the following action:

As it develops its Enterprise Modeling Consortium and other enterprisewide systems for tracking M&O contractor human capital performance metrics, NNSA should consider developing standardized definitions across the enterprise, especially across M&O contractors, to ensure they gather consistent data using human capital metrics with consistent, uniform definitions.

AGENCY COMMENTS AND OUR EVALUATION

We provided NNSA with a draft of this report for their review and comment. NNSA provided written comments. NNSA stated that it appreciated GAO's recognition of the significant challenges NNSA faces in sustaining critical skills in its workforce and the efforts NNSA is taking to identify critical human capital skills, competencies, and levels needed to maintain the nation's nuclear weapons strategy. In addition, NNSA stated that it agreed with the GAO's recommendation that NNSA should consider developing standardized definitions for human capital metrics across the enterprise to help ensure consistent and comparable data. NNSA also provided other additional technical information, which we incorporated where appropriate.

Gene Aloise
Director
Natural Resources and Environment

End Notes

[1] NNSA was created in 1999 under Title 32 of the National Defense Authorization Act for Fiscal Year 2000, Pub. L No. 106-65, § 3201 *et seq.*

[2] M&O contracts are agreements under which the government contracts for the operation, maintenance, or support, on its behalf, of a government-owned or -controlled research, development, special production, or testing establishment wholly or principally devoted to one or more of the major programs of the contracting federal agency. Federal Acquisition Regulation, 48 C.F.R. § 17.601.

[3] The moratorium was begun pursuant to the Energy and Water Development Appropriations Act, 1993, Pub. L. No. 102-377, § 507 (1992). The Comprehensive Nuclear Test Ban Treaty, which would ban all nuclear explosions, was signed by the United States in September 1996 but has not been ratified by the U.S. Senate.

[4] In 1997, the National Defense Authorization Act, Pub. L. No. 104-201 §3162 (1996) established the Commission on Maintaining U.S. Nuclear Weapons Expertise (known as the "Chiles Commission") and directed it to assess and report to Congress on, among other things, DOE's ongoing efforts to recruit and retain scientific, engineering, and technical personnel. The Commission's 1999 report projected that large numbers of retirements over the next few years could further erode the experience and expertise of the critically skilled workforce. See *Report of the Commission on Maintaining U.S. Nuclear Weapons Expertise: Report to the Congress and Secretary of Energy* (Washington, D.C.: Mar. 1, 1999).

[5] GAO, *National Nuclear Security Administration: Contractors' Strategies to Recruit and Retain Critically Skilled Workforce Are Generally Effective*, GAO-05-164 (Feb. 2, 2005).

[6] GAO, *Nuclear Weapons: NNSA Needs More Comprehensive Infrastructure and Workforce Data to Improve Enterprise Decision-making*, GAO-11-188 (Feb. 14, 2011).

[7] The U.S. Senate passed the resolution of ratification for the New Start Treaty on December 22, 2010. President Obama signed the instrument of ratification on February 2, 2011.

[8] GAO, *Government Financial Vulnerability: 14 Areas Needing Special Review*, OCG-90-1 (Washington, D.C.: Jan. 23, 1990).

[9] GAO, *High Risk Series: An Update*, GAO-11-278 (Feb. 16, 2011).

[10] Succession planning is an ongoing, strategic process for identifying and developing a diverse pool of talent for an organization's potential future leaders. GAO, *Diversity Management: Expert-Identified Leading Practices and Agency Examples*, GAO-05-90 (Jan. 14, 2005).

[11] DOE O 426.1, *Federal Technical Capability,* Sept. 20, 2011.

[12] Pub. L. No. 95-91 §621(d) (1977).

[13] Pub. L. No. 103-337 §3161 (1994).

[14] Pub. L. No. 106-65 §3241 (2000).

[15] For excepted service positions, each agency develops, within basic requirements prescribed by law or regulation, its own hiring system, which establishes the evaluation criteria to be used in filling these excepted positions.

[16] DOE O 350.1, Contractor Human Resource Management Programs, Sept. 30, 1996.

[17] GAO-11-188.

[18] GAO, *The Government Performance and Results Act: 1997 Governmentwide Implementation Will Be Uneven*, GAO/GGD-97-109 (Washington, D.C.: June 2, 1997).

[19] To work with weapons, all candidates must be able to obtain a Q-level clearance, which has similar requirements to a top secret clearance in other defense-related agencies.

In: National Nuclear Security Administration ... ISBN: 978-1-62808-190-9
Editors: A. Wright and C. Lewis

Chapter 3

MODERNIZING THE NUCLEAR SECURITY ENTERPRISE: NNSA'S REVIEWS OF BUDGET ESTIMATES AND DECISIONS ON RESOURCE TRADE-OFFS NEED STRENGTHENING*

United States Government Accountability Office

WHY GAO DID THIS STUDY

NNSA, a semiautonomous agency within DOE, is responsible for the nation's nuclear weapons, nonproliferation, and naval reactors programs. Since its inception in 2000, the agency has faced challenges in its ability to accurately identify the costs of major projects. In addition, both the DOE Inspector General, in 2003, and GAO, in 2007, reported concerns with NNSA's PPBE process, specifically in how NNSA validates budget estimates and decides on resource allocations or trade-offs.

GAO was asked to review how NNSA manages programming and budgeting through its PPBE process. GAO examined (1) the current structure of NNSA's PPBE process, (2) the extent to which NNSA reviews its budget estimates, and (3) how NNSA decides on resource trade-offs in its PPBE process. To carry out its work, GAO reviewed NNSA policies, instructions,

* This is an edited, reformatted and augmented version of United States Government Accountability Office, Publication No. GAO-12-806, dated July 2012.

guidance, and internal reports documenting the agency's PPBE process and interviewed NNSA, DOE, and M&O contractor officials.

WHAT GAO RECOMMENDS

GAO recommends that, among other things, DOE update the departmental order for budget reviews, improve the formal process for reviewing budget estimates, and reinstitute an independent analytical capability. The agency agreed in principle with six recommendations but not with one to consolidate various integrated priority lists. GAO continues to believe this recommendation has merit as discussed in the report.

WHAT GAO FOUND

The National Nuclear Security Administration's (NNSA) planning, programming, budgeting, and evaluation (PPBE) process provides a framework for the agency to plan, prioritize, fund, and evaluate its program activities. Formal policies guide NNSA and management and operating (M&O) contractors through each of four phases of the agency's PPBE cycle—planning, programming, budgeting, and evaluation. These phases appear to be sequential, but the process is continuous and concurrent because of the amount of time required to develop priorities and review resource requirements, with at least two phases ongoing at any time.

NNSA does not thoroughly review budget estimates before it incorporates them into its proposed annual budget. Instead, NNSA relies on informal, undocumented reviews of such estimates and its own budget validation review process—the formal process for assessing budget estimates. Neither of these processes adheres to Department of Energy (DOE) Order 130.1, which defines departmental provisions for the thoroughness, timing, and documentation of budget reviews. NNSA officials said the agency does not follow the order because it expired in 2003. Nevertheless, the order is listed as current on DOE's website, and a senior DOE budget official confirmed that it remains in effect, although it is outdated in terminology and organizational structure. Additionally, according to NNSA officials, the agency's trust in its contractors minimizes the need for formal review of its budget estimates. GAO identified three key problems in NNSA's budget validation review process. First, this process does not inform NNSA, DOE, Office of Management and Budget, or congressional budget development decisions because it occurs too late in the budget cycle— after the submission of the President's budget to Congress. Second, this process is not sufficiently thorough to ensure the credibility and reliability of NNSA's budget because it is limited to assessing the processes used to develop budget estimates rather than the accuracy of the resulting estimates and is conducted for a small portion of NNSA's budget—approximately 1.5 percent of which received such review in 2011. Third, other weaknesses in this process, such as no formal evaluative mechanism to determine if corrective actions were taken in response to previous findings, limit the process's effectiveness in assessing NNSA's budget estimates.

NNSA uses a variety of management tools to decide on resource trade-offs during the programming phase of the PPBE process. One of these tools, integrated priority lists—which rank program activities according to their importance for meeting mission requirements—is to provide senior managers with an understanding of how various funding scenarios would affect program activities. However, NNSA has weakened its ability to gauge the effects of resource trade-offs. For example, in 2010, NNSA disbanded its Office of Integration and Assessments, created in response to DOE Inspector General and GAO recommendations that NNSA establish an independent analysis unit to perform such functions as reviewing proposals for program activities and verifying cost estimates. NNSA agreed with these recommendations and, in 2009, instituted the office to identify, analyze, and assess options for deciding on resource trade-offs. Without an independent analytical capability, NNSA may have difficulty making the best decisions about what activities to fund and whether they are affordable.

ABBREVIATIONS

AFR	Administrator's Final Recommendation
DOD	Department of Defense
DOE	Department of Energy
M&O	management and operating
NNSA	National Nuclear Security Administration
OMB	Office of Management and Budget
PPBE	planning, programming, budgeting, and evaluation

July 31, 2012

The Honorable Dianne Feinstein
Chairman
The Honorable Lamar Alexander
Ranking Member
Subcommittee on Energy and Water Development
Committee on Appropriations
United States Senate

The National Nuclear Security Administration (NNSA)—a semiautonomous agency within the Department of Energy (DOE)— manages the nation's nuclear security programs, the primary goals of which are to maintain the safety, security, and effectiveness of the U.S. nuclear weapons stockpile; reduce the global threat posed by nuclear proliferation and terrorism; and provide nuclear propulsion systems for the U.S. Navy. To manage these programs, NNSA oversees eight separate sites—collectively known as the nuclear security enterprise—that are managed and operated by private contractors, known as management and operating (M&O) contractors.[1] Among other things, these contractors operate and maintain the government-owned facilities and infrastructure deemed necessary to support the nuclear security enterprise and to support the capabilities to conduct scientific, technical, engineering, and production activities. To fund these activities, the President requested more than $11.5 billion for NNSA in his fiscal year 2013 budget submission to Congress. Sound management practices, such as linking program activities to defined objectives and ensuring the reliability of budget estimates, can help NNSA use agency resources wisely. It is particularly important for NNSA to embrace this approach because, as our past work has shown, it has faced numerous challenges in accurately estimating the costs of major projects since its inception in 2000. For example, we reported in March of this year that the estimated cost to construct the Chemistry and Metallurgy Research Replacement Nuclear Facility at NNSA's Los Alamos National Laboratory rose from between $745 million and $975 million in 2005 to between $3.7 billion and $5.8 billion in 2010—a near sixfold increase.[2] In addition, in February of this year we testified that the estimated cost to construct the modern Uranium Processing Facility at NNSA's Y-12 National Security Complex jumped from between $600 million and $1.1 billion in 2004 to between $4.2 billion and $6.5 billion in 2011—a near sevenfold increase.[3] Furthermore, in 2000 and 2009, respectively, we reported that NNSA's efforts to extend the operational lives of nuclear weapons experienced a $300 million cost increase for the refurbishment of one warhead and a $70 million increase for another.[4] Given NNSA's record of weak management of major projects, we believe improved federal oversight of NNSA's modernization of the nuclear security enterprise is critical to ensuring that resources are spent in as an effective and efficient manner as possible.

Under section 3252 of the National Defense Authorization Act for Fiscal Year 2000—the NNSA Act—NNSA was required to develop a planning, programming, and budgeting process that operated under sound financial and fiscal management principles. 5 The NNSA Act also requires NNSA to annually submit to Congress a Future Years Nuclear Security Program plan that details NNSA's planned expenditures for the next 5 years. The value of this plan depends, in part, on the accuracy of the cost estimates supporting each program. A realistic cost estimate provides a basis for accurate budgeting and effective resource allocation, which increases the probability of a program's success in meeting its goals. NNSA's first Administrator indicated that he intended to comply with the NNSA Act by instituting a planning, programming, budgeting, and evaluation (PPBE) process—which uses short- and long-term planning to define program priorities and match them to available budgetary resources—that is similar to the Department of Defense's (DOD) budgetary process. As we have reported, DOD's PPBE process logically attempts to plan programs and develop budgets based on high-level strategic goals, but at times it has been poorly implemented.[6] For example, DOD decision makers often commit to more programs than the department can reasonably support. Nonetheless, DOD's process has the potential to provide comparisons of the cost-effectiveness of program activities and to help managers develop the kinds of detailed program and budget plans called for in the NNSA Act. However, both the DOE Inspector General and GAO have reported on concerns with the development and implementation of NNSA's PPBE process.[7] Specifically, these reports identified deficiencies in how NNSA ensures the validity of its budget estimates and how it decides to allocate its resources (resource trade-offs). In our January 2007 report, we recommended, among other things, that NNSA complete all of its PPBE policy guidance. In response, NNSA issued PPBE policy in the form of policy letters that identify the responsibilities of NNSA management, program and site offices, and contractors throughout the agency's PPBE cycle.

In this context, you asked us to determine how NNSA manages its programming and budgeting activities through its PPBE process. For this report, our objectives were to examine (1) the current structure of NNSA's

PPBE process, (2) the extent to which NNSA reviews its budget estimates, and (3) how NNSA decides on resource trade-offs in its PPBE process.

To address these objectives, we analyzed NNSA documentation—including policies, instructions, guidance, and internal reports—related to the implementation of NNSA's PPBE policy across NNSA, as well as program-specific documentation from the Offices of Defense Programs, Defense Nuclear Nonproliferation, and Naval Reactors; these three offices correspond to NNSA's primary missions and collectively account for approximately 85 percent of the President's fiscal year 2013 NNSA budget submission to Congress.[8] To examine the current structure of NNSA's PPBE process, we reviewed NNSA policy documents, guidance, and instructions. To determine the extent to which NNSA reviews its budget estimates, we reviewed NNSA's formal budget validation review process and associated internal NNSA reports and met with officials involved in the development, oversight, or execution of NNSA budget estimate reviews. To identify the tools NNSA uses to decide on resource trade-offs, we analyzed how NNSA determines which program activities it considers necessary to achieve NNSA goals. Additionally, to address all these objectives, we met with officials from the NNSA Offices of the Administrator, Management and Budget, Defense Programs, Defense Nuclear Nonproliferation, Naval Reactors, and Acquisition and Project Management; the site offices for Los Alamos, Sandia, and the Y-12 National Security Complex; and the Naval Reactors Laboratory Field Office; DOE officials from the Office of Budget; and M&O contractor officials from Los Alamos and Sandia National Laboratories, the Y-12 National Security Complex, and Bettis Atomic Power Laboratory.[9] Because NNSA's Office of Naval Reactors is organized as a separate entity within NNSA reporting both to NNSA and the U.S. Navy, we also met with Navy officials from its offices of Financial Management and Budgeting, and Cost Engineering and Industrial Analysis. Appendix I contains a detailed description of our objectives, scope and methodology.

We conducted this performance audit from July 2011 to July 2012 in accordance with generally accepted government auditing standards. Those standards require that we plan and perform the audit to obtain sufficient, appropriate evidence to provide a reasonable basis for our findings and conclusions based on our audit objectives. We believe that the evidence obtained provides a reasonable basis for our findings and conclusions based on our audit objectives.

BACKGROUND

NNSA oversees programs to carry out its missions for nuclear weapons, nonproliferation, and naval nuclear propulsion, for which the President's budget submission to Congress requested more than $11.5 billion for fiscal year 2013—about 42 percent of DOE's total budget. NNSA has primary mission responsibilities in three areas. First, it is responsible for providing the United States with safe, secure, and reliable nuclear weapons in the absence of underground nuclear testing and maintains core competencies in nuclear weapons science, technology, and engineering. Second, NNSA implements key U.S. government nuclear security, nonproliferation, and arms control activities, including securing vulnerable nuclear and radiological material at facilities throughout the world; removing plutonium and highly enriched uranium from partner countries; eliminating U.S. nuclear material declared surplus to defense needs; negotiating and providing the technical capability to verify arms control treaties and agreements; strengthening other countries' capacities to implement nonproliferation obligations; and enhancing other nations' capabilities to deter and detect illicit movement of nuclear and radiological materials. Third, NNSA provides the research, development, design, and operational support for militarily effective naval nuclear propulsion plants, as well as enriched uranium for fabrication into fuel for the Navy's propulsion reactors. NNSA receives four congressional appropriations to fund its activities, three of which align with its primary missions—Weapons Activities (for Defense Programs), Defense Nuclear Nonproliferation, Naval Reactors—and one that funds its management activities—Office of the Administrator.[10]

Since the Manhattan Project produced the first atomic bomb during World War II, NNSA, DOE, and predecessor agencies have depended on the expertise of private firms, universities, and others to carry out research and development work and efficiently operate the government-owned, contractor-operated facilities necessary for the nation's nuclear defense. NNSA conducts its activities at research and development laboratories, production plants, and other facilities (collectively referred to as the nuclear security enterprise). Specifically, NNSA operates three national laboratories that design and ensure the reliability of nuclear weapons— Lawrence Livermore National Laboratory, California; Los Alamos National Laboratory, New Mexico; and the Sandia National Laboratories, New Mexico and California; and four nuclear weapons production sites—the Pantex Plant, Texas; the Y-12 National Security Complex, Tennessee; the Kansas City Plant, Missouri; and the

Savannah River Site, South Carolina; as well as the Nevada National Security Site.

NNSA's relationship with its contractors has been formally established over the years through its M&O contracts—contracting strategies that give these contractors responsibility to carry out major portions of NNSA's missions and apply their scientific, technical, and management expertise.[11] M&O contractors at NNSA sites operate under NNSA's direction and oversight but largely independently of one another. Various headquarters organizations within NNSA develop policies and NNSA site offices, colocated with NNSA's sites, conduct day-to-day oversight of the M&O contractors, and evaluate the M&O contractors' performance in carrying out the sites' missions.

NNSA HAS ESTABLISHED A FORMAL PPBE PROCESS THAT INCLUDES FOUR DEFINED PHASES

According to NNSA's policy, the NNSA PPBE process is composed of four phases—planning, programming, budgeting, and evaluation—and their associated activities. The different phases of PPBE appear sequential, but because of the amount of time required to develop and review resource requirements, the process is continuous and concurrent with at least two phases ongoing at any given time, including phases for different fiscal years. Figure 1 shows the four phases and the months during which each phase is scheduled to occur, according to NNSA policies and guidance.

Planning. According to NNSA policy, in this phase, scheduled to begin annually in November, NNSA is to identify the goals it needs to achieve over the next 5 years and the program activities needed to meet these goals. According to NNSA officials, these goals are defined in a variety of documents, including presidential directives, policy statements, and DOE and NNSA strategic plans. This phase begins with the issuance of NNSA's annual Strategic Planning Guidance, which provides any updates to the strategic plans and identifies any emerging issues. The NNSA program offices use this guidance to conduct their own internal planning processes, update their multiyear plans, including revising or adding program activities needed to meet the agency's goals.

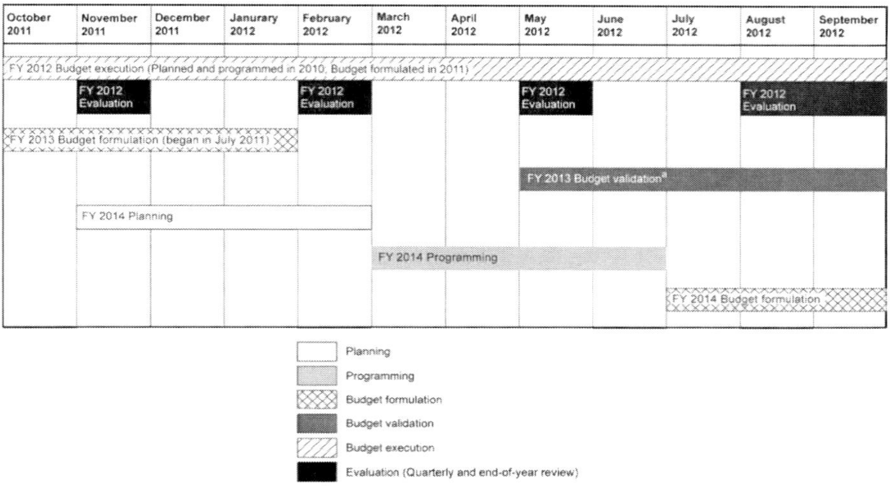

Source: GAO analysis of NNSA documentation.

[a]NNSA policy for budget validation calls for validation activities to begin in the April-May time frame and continue to September each year. However, in practice, the schedule of these activities can vary from year to year. For example, the NNSA memorandum initiating the budget validation process in 2011 scheduled validation activities to occur from June to October of that year.

Figure 1. NNSA PPBE Schedule.

Programming. According to NNSA policy, in this phase, scheduled to begin annually in February, NNSA is to determine which program activities and funding levels it will include in its budget proposal to DOE for the fiscal year beginning in October of the following calendar year. This determination is based on analysis of the activities' estimated costs, as well as the need to meet the NNSA goals defined in the planning process. To determine these activities, NNSA program offices are to work with their contractors to obtain estimates for the cost of the program activities identified in the planning phase and determine how to accomplish these activities within anticipated funding levels, which are defined in annual NNSA Program and Fiscal Guidance. NNSA program offices are to then rank these activities in order of priority for meeting program goals and document these decisions in integrated priority lists. These lists can include proposed program activities above the anticipated funding levels specified in NNSA guidance—these proposed activities are known as unfunded requirements. Using these lists, as well as other briefing materials, a group of senior NNSA officials including the heads of all program offices—the Program Review Council—then is to meet with the Principal

Deputy Administrator to discuss and defend each program's proposed program activities. After reviewing the deliberations of the Program Review Council and the associated documents provided by the program offices, the NNSA Administrator is to decide on resources tradeoffs that result in the combination of program activities that best meet NNSA's goals over the 5-year period covered by the Future Years Nuclear Security Program plan. The Administrator is responsible for issuing the Administrator's Final Recommendations (AFR), scheduled to be completed in May at the conclusion of the programming phase, to document NNSA's justification for its priorities and to serve as the basis for the agency's participation in DOE's program review process, the Strategic Resources Review.

Budgeting. According to NNSA policy, this phase is to integrate NNSA planning and programming priorities and budget estimates into DOE's departmental budget process and consists of the following three parts:

- *Budget formulation*. During formulation, which is scheduled to begin annually in July for the fiscal year beginning in October of the following calendar year, NNSA submits its proposed budget to DOE and participates in the Strategic Resources Review. If DOE's budget deliberations result in changes to NNSA's proposed budget, NNSA may have to rebalance its work activities. In September each year, DOE submits its proposed budget to the Office of Management and Budget (OMB) for review. Depending on OMB revisions, NNSA may need to again revise its work activities. These revisions are incorporated into the President's final budget request for DOE, which is submitted to Congress in February.
- *Budget validation*. According to NNSA guidance, the agency uses its budget validation review process to ensure its budget request is consistent with NNSA priorities and its budget estimating processes are reasonable. NNSA policy calls for NNSA's Office of PPBE to manage a three-phase process of validating approximately 20 percent of NNSA's programs each year, so that 100 percent of its budget is validated every 5 years.[12] Programs to undergo validations are to be determined by a combination of factors, including Program Managers' request, Administrator direction, and/or significant external interest/high program visibility. During Phase I of the process, scheduled for completion in June, before the beginning of the fiscal year in October, program officials determine if their activities conform with strategic guidance and program plans and review their

methods for formulating budgets. In Phase II, scheduled annually for July to September, NNSA contractors or program offices, whichever developed the budget estimates, conduct a self-assessment of their budget planning, formulation, and cost-estimating processes. Phase II self-assessments are to be reviewed by a team—known as a validation review team—that comprises NNSA headquarters and site office staff. During Phase III, scheduled to occur from July through August, these validation review teams also review the cost-estimating practices used by the NNSA contractors and program offices. Importantly, NNSA's validation guidance emphasizes that reviews should focus on the processes used to formulate budget plans and derive budget estimates rather than on the accuracy of the resulting estimates. According to this guidance, validation review teams are to issue a report on their findings on Phases II and III in September to inform NNSA, DOE, and OMB decisions for the following year's budget cycle.

- *Budget execution.* According to NNSA policy, during this process, DOE and NNSA are to allocate, distribute, and control funds to achieve the priorities established in the programming phase, and to maintain the fiscal limits set up in the budgeting phase, which are subject to appropriation of funds by Congress. The execution coincides with the fiscal year and commences once appropriated funds become available—whether by appropriation or Continuing Resolution—at the beginning of the fiscal year every October.

Evaluation. According to NNSA policy, NNSA is to employ an ongoing cycle of evaluations to review program performance. Evaluations are to include annual and quarterly NNSA performance reviews, performance reviews conducted as part of the Government Performance and Results Act, reviews conducted by OMB, and DOE oversight activities.

NNSA Does Not Thoroughly Review Budget Estimates When Developing Its Annual Budget

NNSA does not thoroughly review budget estimates before it incorporates them into its annual budget request. Instead, it relies on undocumented, informal reviews of these estimates by site and headquarters program office officials and the formal budget validation reviews, which conclude after the submission of the President's budget to Congress. Neither of these processes meets DOE provisions for ensuring the credibility and reliability of agency budgets, as defined in DOE Order 130.1. According to senior NNSA officials, NNSA does not comply with DOE Order 130.1 because it believes the order expired in 2003 and therefore no longer applies to NNSA budget activities. Furthermore, they stated that the need for a formal review of budget estimates is minimized, in part, because of the inherent trust between NNSA and its M&O contractors. Additionally, we identified three key problems in NNSA's formal budget validation review process: it occurs too late to affect budget decisions, is not sufficiently thorough, and includes other weaknesses that limit its effectiveness.

NNSA's Process for Reviewing Budget Estimates Is Not Thorough or Documented

NNSA does not have a thorough, documented process for assessing the validity of its budget estimates prior to their inclusion in the President's budget submission to Congress, thereby limiting the reliability and credibility of the budget submission. Specifically, according to NNSA officials from NNSA's Offices of Management and Budget, Defense Programs; Defense Nuclear Nonproliferation; and the site offices for Los Alamos, Sandia, and Y-12, during the programming phase of PPBE, site and headquarters program office officials conduct informal, undocumented reviews of the budget estimates that M&O contractors submitted to determine their reasonableness, though some officials noted that the level of review may vary across site and headquarters program offices.[13] According to these officials, this informal review is often conducted by comparing current budget estimates with those from previous years because the work is largely the same from year to year. If the estimates are similar, and no major programmatic change has taken place, site office and headquarters program office officials said that they generally view these budget estimates as reasonable for inclusion in NNSA's budget estimate. However, site office officials told us that their ability to thoroughly review budget estimates is limited. For example, according to NNSA officials at the Los Alamos Site Office, they do not have the personnel needed or the time, because of other laboratory management responsibilities, to oversee the laboratory's budget estimation practices. They told us that only one dedicated budget analyst is employed at the site office and, because of insufficient personnel resources in the office, a majority of this analyst's time is spent conducting work that is not directly related to budget oversight. NNSA officials from the Y-12 Site Office also told us that they informally review budget estimates when they initially submit them to headquarters program offices. However, they also stated that they become more involved in reviewing budget estimates when the agency is formulating its final budget submission, and the M&O contractors are asked to develop multiple iterations of budget estimates based on various hypothetical funding scenarios. However, these officials also stated that their reviews are not documented. NNSA officials from Defense Programs' Office of Analysis and Evaluation told us that the presence of certified cost engineers—individuals with professional certification in the field of cost assessment and program management—at the NNSA site offices could enhance NNSA's ability to understand how M&O contractors and programs develop budget estimates and assess those estimates.

The practices the site and headquarters program offices follow do not align with the criteria for thoroughness or documentation established in DOE Order 130.1. Specifically, DOE Order 130.1 states that contractor-developed budget estimates should be thoroughly reviewed and deemed reasonable prior to their inclusion in agency budgets and that these reviews should be documented. Senior officials from NNSA's Office of Management and Budget told us that the agency does not strictly adhere to DOE Order 130.1 because it believes that the order has expired and no longer applies to NNSA budget activities. According to these officials, this order expired in 2003, and they are unaware of any other DOE or federal government requirement to conduct budget validation reviews.[14] They further stated that NNSA is conducting budget validation reviews only because it considers them to be a good business practice and that NNSA will work with DOE on updating the order if DOE initiates that process. NNSA officials stated that, if DOE updated and reissued DOE Order 130.1, it would comply to the extent that it had the resources to do so. However, DOE Order 130.1 remains listed on DOE's "All Current Directives" website, and a senior DOE budget official told us that DOE Order 130.1 remains an active order. Additionally, this official stated that a key principle of DOE Order 130.1—federal oversight of contractors' practices for budget formulation—is appropriate and valid. This official noted, however, that the order is outdated in terms of the terminology it uses to describe DOE—it was issued in 1995, predating the 2000 establishment of NNSA—and should be updated to reflect the department's current organizational structure. Furthermore, in March 2009, we issued a cost-estimating guide—a compilation of cost-estimating best practices drawn from across industry and government—in which we reported that validation is considered a best practice to ensure that cost data are credible and reliable for use in justifying estimates to agency management.[15] As a result, NNSA's site and headquarters program office reviews of budget estimates are neither thorough nor documented. According to the Principal Deputy Administrator, NNSA continues to face challenges in moving away from its historical process for developing budgets based solely on the unreviewed estimates produced by NNSA M&O contractors and that NNSA's practices for understanding its program activity costs are not as sufficient as they need to be.

In contrast, NNSA's Office of Naval Reactors' is jointly staffed and funded by both NNSA and the Navy and is therefore subject to naval and DOD, as well as NNSA, standards for reviewing contractor-developed budget estimates. The Office of Naval Reactors conducts a semiannual process—known as budget confirmation—to review all contractor-developed budget estimates. This review is conducted and documented by NNSA technical experts and approved by the Director of the Office of Naval Reactors; this director manages both NNSA and the Navy's activities within the office and has final budgetary decision authority. Officials in NNSA's Office of Management and Budget told us that the Office of Naval Reactors' process is much more rigorous than that used by other NNSA program offices we reviewed. Furthermore, NNSA has exempted the Office of Naval Reactors from NNSA's formal budget validation review process because of management's confidence in the quality of the office's budget confirmation process.

Senior officials in NNSA's Office of Management and Budget told us that NNSA does not have the financial and personnel resources needed to conduct budget estimate reviews with the same rigor as the Navy and DOD. Furthermore, these officials said, the need for a formal review of M&O contractor-developed budget estimates is minimized within NNSA because site office officials have historical knowledge of work with NNSA's M&O contractors that allows them to assess the reasonableness of M&O contractor-developed budget estimates without conducting a formal review and because of the "inherent trust" between NNSA and its M&O contractors that results from its contracting strategy with them. Specifically, one of these officials stated that, to a large extent, only the M&O contractors are in a position to know the scientific and engineering details of nuclear weapons and the associated work scope and funding necessary to ensure their safety and reliability. However, for the last 10 years, we have reported that NNSA has significant weaknesses in its ability to control costs and effectively manage its

M&O contractors.[16] We are therefore concerned that NNSA management continues to deny the need for NNSA to improve its processes for developing credible and reliable budget estimates.

NNSA's Formal Budget Validation Review Process Occurs Too Late to Affect Budget Decisions and Is Not Sufficiently Thorough

We identified three key problems in NNSA's annual budget validation review process—its formal process for assessing M&O contractor- and program-developed budget estimates. First, NNSA's annual budget validation review process occurs too late in the budget cycle to inform NNSA, DOE, OMB, and congressional budget development or appropriations decisions. DOE Order 130.1, which is referenced in NNSA's policy for its budget validation review process, states that agencies should thoroughly review budget estimates before using these estimates to develop budgets. However, NNSA's Phase II and Phase III budget validation reviews are scheduled to begin 5 months after the President submits his budget to Congress. Additionally, during each of the past four budget validation cycles, NNSA did not complete its budget validation reports for at least 12 months following the President's budget submission to Congress and at least 4 months after the beginning of the fiscal year for which NNSA reviewed the budget estimates. Therefore, Congress considered the budget request for NNSA and appropriated funds to it, and NNSA executed these funds to M&O contractors, before NNSA had published the results of the budget validation reviews. Because of their timing, NNSA's budget validation reviews cannot inform NNSA's budget development, DOE or OMB reviews, or Congress' appropriation processes. According to NNSA policy, the timing of NNSA's budget validation review process is designed to inform the NNSA, DOE, and OMB budgeting processes for the fiscal year following that for which the budget validation reviews were conducted. However, the timing of the publication of the budget validation review reports for each of the last 4 years precluded even such delayed consideration because they were issued following the OMB budget formulation process for the following fiscal year.

Second, NNSA's budget validation review process is not sufficiently thorough to ensure the credibility and reliability of NNSA's budget. DOE Order 130.1 states that budgets should be based on budget estimates that have been thoroughly reviewed by site and headquarters program offices. However,

NNSA's budget validation review process is limited to assessing the processes M&O contractors and programs used to develop budget estimates rather than the accuracy of the resulting budget estimates.[17] NNSA's 2010 budget validation review guidance states that the agency lacks the resources and expertise needed to thoroughly evaluate the accuracy of budget estimates on its own and therefore relies on assessments of the reasonableness of the processes used by M&O contractors to develop budget estimates. NNSA officials from the Los Alamos and Y-12 Site Offices told us that they believe the budget validation review process would benefit NNSA more if it more thoroughly assessed the budgetary processes M&O contractors used to develop their budget estimates. Furthermore, NNSA policy and budget validation review guidance stipulate that 20 percent of the agency's programs should be reviewed annually to help ensure its validity, but NNSA's formal validation process actually results in significantly smaller portion of its budget being reviewed. For example, in 2011, NNSA's annual budget validation guidance identified four programs subject to budget validation review—the Engineering Campaign, Nuclear Counterterrorism Incident Response, Global Threat Reduction Initiative, and Fissile Materials Disposition—each of which is conducted at multiple NNSA sites.[18] However, NNSA conducted validation reviews at only one site for each of these programs, which resulted in a formal validation review of approximately 12, 21, 15, and 4 percent of each of those programs' total budgets, respectively, which, together, constituted 1.5 percent of NNSA's budget request for fiscal year 2012.

Third, other weaknesses in NNSA's budget validation review process limit its effectiveness as a resource to assess the validity of its budget estimates. In particular, NNSA workgroups that reviewed the 2007 and 2008 budget validation review cycles recommended that NNSA formally evaluate the status of recommendations made during previous budget validation reviews. However, NNSA has not incorporated a formal mechanism for implementing an evaluation into its budget validation review process. NNSA officials at the Los Alamos and Y-12 site offices also told us that not having such an evaluative mechanism was a weakness in NNSA's budget validation process. Without a formal mechanism, NNSA is limited in its ability to measure (1) any progress M&O contractors or programs have made in their processes for estimating budgets in response to recommendations from previous budget validation reviews and (2) the effectiveness of NNSA's budget validation review process. For example, a 2010 budget validation review of the Readiness Campaign recommended that the program more formally document its budget processes, guidance, and estimating assumptions.[19] Furthermore, a 2009 budget validation review of the Elimination of Weapons Grade Plutonium Production program found that the program could not provide documentation of its internal budget processes.[20] However, in both instances, NNSA did not follow up to determine if the programs had addressed these concerns during subsequent budget validation reviews. Additionally, budget validation reviews do not always include recommendations to improve M&O contractor or program processes for estimating budgets when they identify potentially serious weaknesses in those M&O contractor's or programs' ability to develop cost estimates. For example, according to a 2010 budget validation review of budget estimation activities for the Nonproliferation and Verification Research and Development program at Sandia National Laboratories, six of the eight projects reviewed lacked sufficient documentation to support their cost estimates, including two that lacked any supporting documentation.[21] The report noted the importance of credible cost estimates, but it did not formally recommend any remedial improvements and rated the overall processes used to develop those cost estimates as satisfactory. Additionally, NNSA officials in the Defense Programs' Office of Analysis and Evaluation told us that the cost information used to support budget validation review reports is often flawed or nonexistent.

NNSA Has Implemented Some Tools to Support Decision Making on Resource Trade-offs but Has Stopped Using or Developing Other Capabilities

During the programming phase of PPBE, NNSA uses a variety of management tools, such as integrated priority lists and requirements and resources assessments, to support its programming phase and assist senior managers in making decisions on resource trade-offs. However, it has stopped using these capabilities or developing others.

NNSA Has Developed and Implemented Some Tools to Assist Management in Deciding on Resource Trade-offs

NNSA uses the following management tools to decide on resource trade-offs during the programming phase of its PPBE process:

Integrated priority lists. NNSA's policy for the programming phase stipulates that each of NNSA's nine program offices is to annually develop an integrated priority list that ranks program activities according to their importance for meeting mission requirements. These lists provide senior NNSA and DOE managers with an understanding of how various funding scenarios would affect program activities. Specifically, these lists rank the priority of program activities that are within anticipated appropriation levels—which are of the highest priority—as well as those that NNSA would fund if the appropriation levels were sufficiently high to do so. For example, the program activity listed last on an integrated priority list would be the first to forgo funding if appropriation levels are lower than anticipated. Conversely, these lists define program activities— unfunded requirements—that would be funded if appropriation levels are higher than anticipated. NNSA instructions for the programming phase stipulate that the agency is to combine the nine program office-developed integrated priority lists into four that correspond to the four congressional appropriations NNSA receives. Three of the integrated priority lists— those of the Offices of the Administrator, Defense Nuclear Nonproliferation, and Naval Reactors—correspond directly to specific appropriations,[22] but NNSA does not combine the six others that represent activities funded by the Weapons Activities appropriation into a single integrated priority list. NNSA officials stated that these six others represent separate and distinct mission areas and only the Administrator can decide on the resource trade-offs among them. Of the six program offices funded by the Weapons Activities appropriation, Defense Programs accounts for a large majority—approximately 82 percent—of the funding requested in NNSA's fiscal year 2013 budget submission to Congress. According to officials in NNSA's Office of Management and Budget, the Administrator is responsible for deciding on how to allocate resources across program offices. However, these decisions are not documented or reflected in a single integrated priority list for program activities funded by the Weapons Activities appropriation. By not combining these lists, NNSA is limiting the formal documentation available to inform DOE about how various Weapons Activities appropriation funding scenarios would affect the program activities conducted by these six program offices. Specifically, because these six lists are not integrated, it is unclear which program activities would be affected by changes to appropriation levels or which programs across the six lists are of the highest priority.

Requirements and Resources Assessments. During the 2010 and 2011 programming cycles, NNSA used its Requirements and Resources Assessment process to independently assess the need to conduct new program activities and unfunded requirements. According to the NNSA handbook for this process, officials in NNSA's Office of Management and Budget review program offices' budget submissions for activities that are either above anticipated funding levels or are for new activities. For these assessments, officials are to analyze specific factors related to these activities, such as their need for meeting agency priorities and the reasonableness of the assumptions used to produce their budget estimates. The objective of this process is to ensure that new program activities and unfunded requirements are needed to meet NNSA priorities. For example, according to officials in NNSA's Office of Management and Budget, the use of the Requirements and Resources Assessment process was a contributing factor in reducing the amount of unfunded program activities included in NNSA's budget from approximately $1 billion for fiscal year 2012 to approximately $80 million for fiscal year 2013. Furthermore, draft NNSA guidance states that the process has identified inconsistencies in the quality of estimates and the level of insight and understanding program managers have regarding the fidelity of the estimates supporting their budgets. According to officials in NNSA's Office of Management and Budget, this process is a simple and effective tool for providing management with additional information on the need to conduct proposed new program activities or unfunded requirements.

However, these officials also stated that this process is time-consuming and would not be practical or efficient to apply to the entirety of NNSA program activities because it was designed to assess program components rather than entire programs; they added that other types of program reviews or validations would be better suited for conducting program needs analysis on an enterprise-wide basis. Additionally, because the NNSA Office of Integration and Assessments, which was responsible for conducting these assessments, was dissolved in 2010, officials in NNSA's Office of Management and Budget told us that they may discontinue the use of Requirements and Resources Assessment process in future programming cycles. Furthermore, in the current austere budget environment, they do not foresee any programs proposing activities that are either new or above anticipated funding levels. Therefore, the continued use of this process in future programming cycles is uncertain. However, we believe that NNSA has demonstrated this process can be an important tool for assessing the necessity to fund certain activities in order to meet its mission requirements.

Enterprise Portfolio Analysis Tool. NNSA's Office of Defense Programs is implementing a data system—the Enterprise Portfolio Analysis Tool—

designed to provide a consistent framework for managing the PPBE process within Defense Programs, which accounts for 54 percent, or $6.2 billion, of the President's $11.5 billion fiscal year 2013 budget request for NNSA. As we testified in February 2012, a tool such as this could help NNSA obtain the basic data it needs to make informed management decisions, determine return on investment, and identify opportunities for cost saving.[23] Currently, this tool includes a mechanism to identify when decisions on resource trade-offs must be made if, for example, M&O contractor-developed budget estimates for program requirements exceed the budget targets NNSA provided for those programs. Additionally, the tool is to incorporate Defense Programs' common work activity structure—known as its work breakdown structure—to facilitate an analysis of consistent budget data from across the NNSA enterprise. Specifically, the tool may allow Defense Programs managers to compare the budget estimates for analogous activities across the nuclear security enterprise regardless of which M&O contractor or program is conducting them. Furthermore, Defense Programs officials stated that they eventually plan to use this tool to compare budget estimates of program activities with the amounts the programs ultimately expended, but they said that the introduction of this capability is not imminent. According to Defense Programs and M&O contractor officials, the implementation of this tool is placing an additional labor burden on NNSA M&O contractors because of the quantity of historical budget data that need to be entered into it. However, according to Defense Programs officials, once these initial historical data are entered, the M&O contractors will need to annually update the system with the most recent year's data.

NNSA Does Not Use Capabilities That Were Previously Implemented or Partially Developed to Support Its Programming Activities

NNSA no longer has an independent analytical capability to perform such functions as reviewing proposals for program activities and verifying cost estimates. In addition, since 2009, the NNSA Administrator has not formally documented his decisions on resource trade-offs at the close of the programming phase in the AFR. Furthermore, NNSA has not completed cost estimating guidance to assist NNSA program managers in identifying reliable M&O contractor practices for estimating costs for operations and sustainment activities. By not using these capabilities, NNSA has reduced its ability to decide on resource trade-offs because it has not enhanced, made formal, or implemented capabilities that it had already, to varying degrees, developed or used.

Independent Analytical Capability

The DOE Inspector General and GAO, recommended in 2003 and 2007, respectively, that NNSA establish an independent analysis unit to perform such functions as reviewing proposals for program activities and verifying cost estimates.[24] NNSA agreed with these recommendations and, in 2009, instituted the Office of Integration and Assessments to identify, analyze, assess, and present to senior NNSA management options for managing its programs and making decisions on resource trade-offs. The specific responsibilities of this office included analyzing program performance, evaluating programming and funding alternatives, and assessing the implementation and effectiveness of process improvement initiatives. Furthermore, this office managed the Requirements and Resources Assessment process during the 2010 programming cycle. However, NNSA disbanded the office in 2010, 18 months after it was formally created. NNSA officials also told us that it was never properly staffed, which thereby limited its effectiveness.

In the memorandum establishing the Office of Integration and Assessments, NNSA stated that it expected the office, in conjunction with DOE's Office of Cost Analysis, to provide DOD-like analytical resources across NNSA.[25] Since then, however, DOE has also eliminated its Office of Cost Analysis. With both of these offices now gone, neither NNSA nor DOE has independent cost assessment or program evaluation capabilities analogous to those of DOD. In contrast, because NNSA's Office of Naval Reactors is organized as a separate entity within NNSA reporting both to NNSA and the U.S. Navy, unlike the rest of NNSA, it is subject to the Navy's independent analytical resources, such as assessments by the Naval Sea Systems Command's Office of Cost Engineering and Industrial Analysis and the Naval Center for Cost Analysis, which conduct independent reviews and analyses of program cost estimates. Furthermore, DOD has an Office of Cost Assessment and Program Evaluation, which has a similar function but with a purview that extends across DOD, including the Navy. These layers of independent review approximate NNSA's vision for independent analysis, as described in the memorandum establishing the Office of Integration and Assessments. Following the dissolution of the Office of Integration and Assessments in 2010, NNSA's Office of Defense Programs created the Office of Analysis and Evaluation to conduct similar program review functions. However, the capabilities of this office are limited by several factors. For example, because the office is positioned within Defense Programs, it does not have purview to conduct analysis on any of NNSA's other programs, which, in total, constitute nearly half of the agency's budget request for fiscal year 2013. Additionally, according to Defense Programs officials, this office does not have the capability to self-initiate reviews of programs but rather is instructed by Defense Programs' management on what activities to assess, thereby limiting the office's independence. Furthermore, NNSA officials from this office stated that properly staffing the office remains a challenge because many qualified individuals left DOE and NNSA when they eliminated the Offices of Cost Analysis and Integration and Assessments, respectively.

Even though NNSA has had difficulty in maintaining an agencywide independent analytical capability, NNSA's Principal Deputy Administrator told us that NNSA remains supportive of the concept of an independent analytical unit to conduct assessments of programs agencywide. However, senior NNSA officials told us that creating and developing the capabilities of such an office would be difficult in the current budget environment and that therefore NNSA has no current plans to institute such a capability.

Administrator's Final Recommendations

The NNSA Administrator has not formally documented his decisions on resource trade-offs at the close of the programming phase in the AFR since 2009, which is inconsistent with NNSA policy and instructions. When issued, this document articulated the Administrator's rationale and methodology for deciding on resource trade-offs during the programming phase of the PPBE process—which one senior official in NNSA's Office of Management and Budget described as an important component of the PPBE process—to support in his budget proposal to DOE and to better facilitate NNSA's participation in DOE's Strategic Resources Review. According to senior NNSA officials, the Administrator considered the AFR to be a useful management tool but decided to discontinue issuing it because of concerns that its contents, which are predecisional Executive Branch deliberative material and embargoed from public release by OMB Circular A-11, could be leaked and thereby reduce the flexibility of DOE and OMB in making final decisions regarding the President's Budget. Instead of the AFR, the Administrator now develops an internal document called "Administrator's Preliminary Decisions," which is not required in NNSA policy, guidance, or instructions; contains more generalized information; and does not have the rationales, methodologies, and justifications for decision making on resource trade-offs that were previously incorporated into the AFR.

Issuance of Supplemental Guide on Cost Estimating

NNSA developed a draft guide—the *Program Managers' Guide to Understanding and Reviewing Cost Estimates for Operations and Sustainment Activities*—in 2010 to assist NNSA program managers in identifying reliable M&O contractor practices for estimating costs for operations and sustainment activities—activities not related to construction; according to this guide, these activities constitute approximately 80 percent of NNSA's annual budget. This guide was also created to supplement the information provided in *NNSA's Business Operating Procedure 50.005, Establishment of an Independent Cost Estimate Policy* and interim *Cost Estimating Guide 50.005*, which identifies best practices for preparing cost estimates.[26] However, the *Program Managers' Guide to Understanding and Reviewing Cost Estimates for Operations and Sustainment Activities* was largely completed but never finalized before NNSA dissolved the Office of Integration and Assessments, which had drafted the guide, and NNSA officials said the agency has no plans to complete or issue it. According to officials in NNSA's Office of Management and Budget, NNSA drafted this guide because it recognized that supplemental information focused on operations and sustainment activities cost estimates—the development of which, according to this guide, are not governed by any specific NNSA guidance or processes—could enhance the tools available to program managers in evaluating cost estimates and how they are translated into budget estimates. The objective of the guide was to provide an instructive document to facilitate program managers' ability to understand what constitutes a rigorous process for ensuring quality operations and sustainment cost estimates on an ongoing basis and evaluate the reasonableness those estimates. This guide also defined key components of cost estimating to clarify the responsibilities and expectations of NNSA program managers and included instructions for how NNSA program managers can assess the quality of budget estimates submitted by M&O contractors. NNSA officials with Defense Programs' Office of Analysis and Evaluation told us that additional guidance on how to assess the costs of operations and sustainment activities could enhance program managers' ability to assess the reliability and credibility of cost estimates.

CONCLUSION

NNSA has established a formal four-phase PPBE process that uses short- and long-term planning to define program priorities and match them to available budgetary resources. However, DOE and NNSA have not taken adequate steps to make this process as effective and efficient as possible. In particular, DOE Order 130.1, which defines DOE's provisions for budget activities, references outdated terminology and organizations that no longer exist within the department, leading to confusion regarding the order's applicability and requirement for implementation. As a result, NNSA believes that the order has expired and that it is not required to adhere to its provisions. By not adhering to these provisions, NNSA is reducing the credibility of its budget proposals. Moreover, NNSA's process for developing budget estimates continues to rely heavily on its M&O contractors to develop budget estimates without an effective, thorough review of the validity of those estimates. Without thorough reviews by site and headquarters program offices of budget estimates, NNSA cannot have a high level of confidence in its budget estimates or in its ability to make informed decisions on resource trade-offs and to enhance the credibility and reliability of its budget. Furthermore, NNSA's formal budget validation review process does not sufficiently ensure the credibility and reliability of NNSA's budget, primarily because of deficiencies in the timing of these reviews. Also, without a formal mechanism to evaluate the status of recommendations made during previous budget validation reviews, NNSA is limited in its ability to measure any progress M&O contractors or programs have made to their budget estimating processes.

NNSA has reduced its ability to decide on resource trade-offs because it has not enhanced, made formal, or implemented capabilities that it has already, to varying degrees, developed or used. In particular, NNSA does not follow its instructions for preparing an integrated priority list for each congressional appropriation, as it does not combine the six priority lists that represent activities funded by the Weapons Activities appropriation into a single integrated list. By not combining these lists into a single integrated priority list, NNSA is limiting the formal documentation available to inform DOE which program activities would be affected by changes to this appropriation. Moreover, NNSA instituted and then disbanded an independent analytic capability that would provide it with an independent cost assessment or evaluation capabilities of the reasonableness and affordability of various programs and projects proposed by NNSA offices. By disbanding its independent analytical capability, NNSA is losing its ability to improve its cost-estimating capabilities and better ensure that its project cost estimates are credible and reliable. Because of the fiscal constraints in the current budget environment, it is all the more critical that NNSA have the capability to conduct independent cost analyses to enhance its ability to make the most effective and efficient resource decisions on resource trade-offs. Despite previous recommendations that DOE's Inspector General made in 2003, and that we made in 2007, to institute an independent analytical capability to assess programs throughout all of NNSA, NNSA continues to lack such a function. Not having this capability could preclude NNSA from making the best decisions about what activities to fund and whether they are affordable. In addition, NNSA may cease using its Requirements and Resources Assessment process—which is intended to provide some independent analysis of new program activities and unfunded requirements—in future PPBE budget cycles because it does not anticipate program proposals for new activities or unfunded requirements. By not retaining this process, NNSA would lose an important tool for assessing the necessity to fund certain activities in order to meet its mission requirements.

Furthermore, NNSA no longer follows its policy to issue the AFR. Without a formal document, NNSA and DOE have no formal record of the Administrator's rationale and methodology for deciding on resource tradeoffs during the programming phase of the PPBE process. We recognize that NNSA needs to hold confidential, internal budgetary and resource trade-off deliberations; however, we do not believe that this need supersedes NNSA policy or the benefits provided by documented decision making during programming, which one senior NNSA official described to us an important component in NNSA's PPBE process. Not issuing the AFR (or some similarly precise documentation) places the Administrator in conflict with official NNSA policy and with an important PPBE precept—the importance of transparency.

Finally, NNSA developed draft guidance in 2010 to assist NNSA program managers in identifying reliable M&O contractor practices for estimating costs for operations and sustainment activities. Such guidance would better equip NNSA program managers to more accurately evaluate the reasonableness of cost estimates, but this guidance is in draft form and NNSA has no plans to complete and issue it. Without such guidance, NNSA program managers are limited in their ability to assess the reliability and credibility of budget estimates.

RECOMMENDATIONS FOR EXECUTIVE ACTION

To enhance NNSA's ability to better ensure the validity of its budget submissions, and to decide on resource trade-offs, we recommend that the Secretary of Energy take the following seven actions:

- Direct the DOE Office of Budget to formally evaluate DOE Order 130.1 and revise as necessary, and communicate any revisions to the NNSA Administrator so that the agency will have updated provisions for assessing the quality of its budget estimates.
- Direct the Administrator of NNSA to:
    - Develop a formal process, or amend its budget validation review process, to ensure that all budget estimates are thoroughly reviewed by site and headquarters program offices, and that these reviews are timed to inform NNSA, DOE, OMB, and congressional budget decisions.
    - Once this process is developed, incorporate a formal mechanism to evaluate the status of recommendations made during previous budget validation reviews so that NNSA can measure M&O contractors' and programs' progress in responding to deficiencies with their budget estimates.
    - Combine the integrated priorities lists for each of the program offices funded within the Weapons Activities appropriation into a single integrated priorities list, as stipulated in NNSA instructions, to better inform DOE which program activities would be affected by changes to this appropriation.
    - Reinstitute an independent analytical capability to provide senior decision makers with independent program reviews, including an analysis of different options for deciding on resource trade-offs, and facilitate NNSA making the best decisions about what activities to fund and whether they are affordable. As part of this capability, formally retain the Requirements and Resources Assessment process to review proposed new activities and unfunded requirements.
    - Reinstitute the issuance of the Administrator's Final Recommendations to document the Administrator's rationale and methodology for deciding on resource trade-offs to support in his budget proposal to DOE and to better facilitate NNSA's participation in DOE's budget process.

- Complete and formally issue the *Program Managers' Guide to Understanding and Reviewing Cost Estimates for Operations and Sustainment Activities* so that program managers will be better equipped to evaluate the reasonableness of cost estimates.

AGENCY COMMENTS AND OUR EVALUATION

We provided DOE with a draft of this report for its review and comment. In its written comments, NNSA, responding on behalf of DOE, provided observations on the report's findings and stated that it generally agreed in principle with six of our seven recommendations and did not concur with one. NNSA did not concur with our recommendation to combine the integrated priorities lists for all program offices funded by the Weapons Activities appropriation into a single integrated priorities list, as is stipulated by NNSA instructions for the programming phase of PPBE.

NNSA agrees that the integrated priorities lists are a useful tool to facilitate NNSA and DOE decision-making. However, NNSA states that it believes reaching management consensus on a single integrated priorities list for these program offices would be a difficult, time consuming process and that its current approach for deciding on resource trade-offs is effective and efficient. We acknowledge that NNSA uses a variety of tools in addition to integrated priorities lists to conduct programming activities, but we continue to believe that combining the integrated priorities lists for all program offices funded by the Weapons Activities appropriation could enhance the agency's ability to support its decisions on resource trade-offs for DOE consideration during the Strategic Resources Review. However, NNSA stated in its comments that it would consider the development of more robust integrated priority lists if circumstances require changes to its current approach.

NNSA further acknowledged that aspects of its PPBE process could be improved but disagreed with our report's characterization of its budget estimate review processes as not being thorough. NNSA commented that it believes that our conclusions overemphasize some procedural areas for potential improvement, without accurately considering the cumulative effectiveness of NNSA's PPBE process as a whole. We continue to believe that the agency's processes for reviewing budget estimates are not sufficiently thorough to ensure the credibility and reliability of those estimates and do not meet the provisions defined in DOE Order 130.1. Specifically, the reviews conducted by site and headquarters program office officials are informal and undocumented, and NNSA's budget validation review process—the agency's formal process for assessing M&O contractor- and program-developed budget estimates—does not assess the accuracy of budget estimates and is conducted for a small portion of the agency's annual budget.

NNSA also provided technical comments, which we incorporated throughout the report as appropriate.

Gene Aloise
Director,
Natural Resources and Environment

APPENDIX I: OBJECTIVES, SCOPE, AND METHODOLOGY

The objectives of our review were to determine (1) the current structure of the National Nuclear Security Administration's (NNSA) planning, programming, budgeting, and evaluation (PPBE) process; (2) the extent to which NNSA reviews its budget estimates; and (3) how NNSA decides on resource trade-offs in its PPBE process.

To determine the current structure of NNSA's PPBE process, we reviewed the NNSA policies and other headquarters-developed instructions and guidance documents that define how the process is designed to function. We also reviewed program-specific PPBE documentation from the Offices of Defense Programs, Defense Nuclear Nonproliferation, and Naval Reactors; these three offices correspond to NNSA's primary missions and collectively account for approximately 85 percent of the President's fiscal year 2013 NNSA budget submission to Congress. We also interviewed officials from NNSA's Office of Management and Budget, which is responsible for managing NNSA's PPBE process, as well as the offices of Defense Programs, Defense Nuclear Nonproliferation, and Naval Reactors to discuss how NNSA's PPBE process is designed to function.

To determine the extent to which NNSA reviews its budget estimates, we reviewed DOE Order 130.1 and NNSA policies, instructions, and guidance that define how such reviews are to be conducted. We also analyzed documentation of the formal budget validation reviews conducted by NNSA for the last five review cycles, as well as the results of two NNSA workgroups that evaluated the budget validation review process. Furthermore, we interviewed officials involved in the development, oversight, or execution of NNSA budget estimate reviews from the NNSA Offices of the Administrator,

Management and Budget, Defense Programs, Defense Nuclear Nonproliferation, Naval Reactors, and Acquisition and Project Management; the site offices for Los Alamos, Sandia, and the Y-12 National Security Complex, and the Naval Reactors Laboratory Field Office; DOE officials from the Office of Budget; and M&O contractor officials from Los Alamos and Sandia National Laboratories, the Y-12 National Security Complex, and Bettis Atomic Power Laboratory.[1] Because NNSA's Office of Naval Reactors is organized as a separate entity within NNSA reporting both to NNSA and the U.S. Navy, we also met with Navy officials from its Offices of Financial Management and Budgeting, and Cost Engineering and Industrial Analysis.

To determine how NNSA decides on resource trade-offs, we reviewed NNSA policies, instructions, and guidance for its programming process. Based on these documents, we identified the tools that NNSA uses, or has used, to assist NNSA management in deciding on resource trade-offs and reviewed documentation of how these tools were applied by program offices and NNSA management during the programming phases of previous PPBE cycles. We also interviewed officials from the NNSA Offices of the Administrator, Management and Budget, Defense Programs, Defense Nuclear Nonproliferation, Naval Reactors to discuss how they decide on, and document, resource trade-offs.

We conducted this performance audit from July 2011 to July 2012 in accordance with generally accepted government auditing standards. Those standards require that we plan and perform the audit to obtain sufficient, appropriate evidence to provide a reasonable basis for our findings and conclusions based on our audit objectives. We believe that the evidence obtained provides a reasonable basis for our findings and conclusions based on our audit objectives.

End Notes

[1] These sites include three national laboratories, four production plants, and one test site.

[2] GAO, *Modernizing the Nuclear Security Enterprise: New Plutonium Research Facility at Los Alamos May Not Meet All Mission Needs*, GAO-12-337 (Washington, D.C.: Mar. 26, 2012).

[3] GAO, *National Nuclear Security Administration: Observations on NNSA's Management and Oversight of the Nuclear Security Enterprise*, GAO-12-473T (Washington, D.C.: Feb. 16, 2012).

[4] GAO, *Nuclear Weapons: Improved Management Needed to Implement Stockpile Stewardship Program Effectively;* GAO-01-48 (Washington, D.C.: Dec. 14, 2000), and GAO, *Nuclear Weapons: NNSA and DOD Need to More Effectively Manage the Stockpile Life Extension Program*, GAO-09-385 (Washington, D.C.: Mar. 2, 2009).

[5] NNSA was created by the National Defense Authorization Act for Fiscal Year 2000 (Pub. L. No. 106-65, § 3201 et seq. [1999]).

[6] GAO, *Defense Acquisitions: A Knowledge-Based Funding Approach Could Improve Major Weapon System Program Outcomes*, GAO-08-619 (Washington, D.C.: July 2, 2008), and GAO, *DOD Weapon Systems: Missed Trade-off Opportunities During Requirements Reviews*, GAO-11-502 (Washington, D.C.: June 16, 2011).

[7] GAO, *National Nuclear Security Administration: Additional Actions Needed to Improve Management of the Nation's Nuclear Programs*, GAO-07-36 (Washington, D.C.: Jan. 19, 2007); and National Nuclear Security Administration's Planning, Programming, Budgeting, and Evaluation Process, DOE/IG-0614, August 2003.

[8] Six other NNSA offices are responsible for carrying out additional program activities: Management and Budget, Emergency Operations, Defense Nuclear Security, Chief Information Officer, Counterterrorism and Counterproliferation, and Acquisition and Project Management.

[9] We selected a nonprobability sample of NNSA sites based on the levels of funding they receive from NNSA for the programs included in the scope of our work. Because we used a nonprobability sample, the results we obtained from these sites are not generalizable across the nuclear security enterprise; however, they did provide us with specific examples of how NNSA conducts PPBE activities at the site level.

[10] The Weapons Activities appropriation also funds the Emergency Operations, Defense Nuclear Security, Chief Information Officer, Counterterrorism and Counterproliferation, and Acquisition and Project Management program offices.

[11] Management and operating contracts are agreements under which the government contracts for the operation, maintenance, or support, on its behalf, of a government-owned or -controlled research, development, special production, or testing establishment wholly or principally devoted to one or more of the major programs of the contracting federal agency. Federal Acquisition Regulation, 48 C.F.R. § 17.601.

[12] NNSA's Office of PPBE is organizationally located within the NNSA Office of Management and Budget.

[13] According to NNSA officials, site office officials conduct these informal reviews of the budget estimates for activities funded by the Weapons Activities appropriation because they are developed by the M&O contractors. Conversely, they stated that the budget estimates for the Defense Nuclear Nonproliferation Appropriation are developed by program federal employees at NNSA Headquarters. For these activities, contractors compete for projects based upon submitted proposals and funding availability.

[14] DOE Notice 251.45 extended DOE Order 130.1 until May 1, 2003.

[15] GAO, *Cost Estimating and Assessment Guide: Best Practices for Developing and Managing Capital Program Costs*, GAO-09-3SP (Washington, D.C.: March 2009).

[16] GAO-01-48, GAO-07-36, and GAO-12-337.

[17] The criteria used by validation review teams to assess the reasonableness of budget estimation processes include completeness, consistency, reasonableness, and documentation.

[18] The Engineering Campaign provides modern components and engineering science capabilities to ensure the safety, security, reliability, and performance of the stockpile. The Nuclear Counterterrorism Incident Response program responds to and mitigates nuclear and radiological incidents worldwide. The Global Threat Reduction Initiative works in the United States and internationally to convert research reactors and medical isotope production processes from the use of highly enriched uranium, remove and dispose of excess nuclear and radiological materials, and protect high-priority nuclear and radiological

sources from theft. The Fissile Materials Disposition program supports the downblending of highly enriched uranium that is excess to U.S. defense needs and is planning to assist Russia in modifying reactors in that country.

[19] The Readiness Campaign identifies, develops, and deploys new or enhanced processes, technologies and capabilities to meet current nuclear weapon design, production, and dismantlement needs and provide quick response to national security requirements.

[20] The Elimination of Weapons Grade Plutonium Production program seeks to facilitate the closure of Russian nuclear reactors by building or refurbishing replacement fossil fuel plants.

[21] The Nonproliferation and Verification Research and Development program's mission is to conduct needs-driven research, development, testing, and evaluation of new technologies that are intended to strengthen the United States' ability to prevent and respond to nuclear, chemical, and biological attacks.

[22] According to NNSA officials, the integrated priority list developed by NNSA's Office of Management and Budget includes all activities funded by the Office of the Administrator appropriation.

[23] GAO-12-473T.

[24] DOE/IG-0614 and GAO-07-36.

[25] DOE established the Office of Cost Analysis in order to improve the department's cost-estimating capabilities and better ensure that its project cost estimates are reliable by providing a new independent cost-estimating function for the department.

[26] The information included in NNSA's interim *Cost Estimating Guide 50.005* is applicable to all cost estimating activities, but one senior official in NNSA's Office of Acquisition and Project Management told us that it is most useful for construction and capital asset projects, which are land, structures, equipment, and information technology used by the federal government and have an estimated useful life of 2 years or more, as well as strategic assets that are unique physical or intellectual property that is of long-term or ongoing value to an enterprise.

End Notes for Appendix I

[1] We selected a nonprobability sample of NNSA sites based on the levels of funding they receive from NNSA for the programs included in the scope of our work. Because we used a nonprobability sample, the results we obtained from these sites are not generalizable across the nuclear security enterprise; however, they did provide us with specific examples of how NNSA conducts PPBE activities at the site level.

In: National Nuclear Security Administration … ISBN: 978-1-62808-190-9
Editors: A. Wright and C. Lewis

Chapter 4

MODERNIZING THE NUCLEAR SECURITY ENTERPRISE: OBSERVATIONS ON DOE'S AND NNSA'S EFFORTS TO ENHANCE OVERSIGHT OF SECURITY, SAFETY, AND PROJECT AND CONTRACT MANAGEMENT*

David C. Trimble

WHY GAO DID THIS STUDY

DOE and NNSA are responsible for managing nuclear weapon- and nonproliferation-related national security activities in national laboratories and other sites and facilities, collectively known as the nuclear security enterprise. Major portions of NNSA's mission are largely carried out by contractors at each site. GAO has designated contract management of major projects (i.e., those $750 million or more) at DOE and NNSA as a high risk area. Progress has been made, but GAO continues to identify security and safety problems at DOE and NNSA sites as well as project and contract management problems related to cost and schedule overruns on major projects.

This testimony addresses DOE's and NNSA's oversight of (1) security performance, (2) safety performance, and (3) project and contract management

---

* This is an edited, reformatted and augmented version of United States Government Accountability Office, Publication No. GAO-13-482T, dated March 13, 2013.

in the nuclear security enterprise. It is based on prior GAO reports issued from August 2000 to December 2012.

DOE and NNSA continue to act on the numerous recommendations GAO has made to improve management of the nuclear security enterprise. GAO will continue to monitor DOE's and NNSA's implementation of these recommendations.

WHAT GAO FOUND

The Department of Energy (DOE) and the National Nuclear Security Administration (NNSA), a separately organized agency within DOE, continue to face challenges in ensuring that oversight of security activities is effective. For example, in July 2012, after three trespassers gained access to the protected security area directly adjacent to one of the nation's most critically important nuclear weapon-related facilities, the Y-12 National Security Complex, DOE and NNSA took a number of immediate actions. These actions included repairing security equipment, reassigning key security personnel, and firing the Y-12 protective force contractor. As GAO and others have reported, DOE has a long history of security breakdowns and an equally long history of instituting remedies to fix these problems. For example, 10 years ago, GAO reported on inconsistencies among NNSA sites on how they assess contractors' security activities and, since that time, DOE has undertaken security initiatives to address these issues. GAO is currently evaluating these security reform initiatives.

DOE and NNSA continue to face challenges in ensuring that oversight of safety performance activities is effective. DOE and NNSA have experienced significant safety problems at their sites, and recent efforts to reform safety protocols and processes have not demonstrated sustained improvements. Long-standing DOE and NNSA management weaknesses have contributed to persistent safety problems at NNSA's national laboratories. For example, in October 2007, GAO reported that nearly 60 serious accidents or near misses had occurred at NNSA's national laboratories since 2000. DOE has undertaken a number of reforms to address persistent safety concerns. For example, in March 2010, the Deputy Secretary of Energy announced a reform effort to revise DOE's safety and security directives. However, GAO reported in September 2012 that DOE's safety reforms did not fully address continuing safety concerns that GAO and others identified in the areas of quality assurance, safety culture, and federal oversight and, in fact, may have actually weakened independent oversight.

DOE and NNSA have made progress but need to make further improvements to their contract and project management efforts. DOE has made progress in managing nonmajor projects—those costing less than $750 million—and in recognition of this progress, GAO narrowed the focus of its high-risk designation of DOE's Office of Environmental Management (EM) and NNSA to major contracts and projects. Specifically, as GAO noted in its December 2012 report on 71 DOE EM and NNSA nonmajor projects, GAO found the use of some sound management practices that were helping ensure successful project completion. However, major projects continue to pose a challenge for DOE and NNSA. For example, in December 2012, GAO reported that the estimated cost to construct the Waste Treatment and Immobilization Plant in Washington State had tripled to $12.3 billion since its inception in 2000, and the scheduled completion date had slipped by nearly a decade to 2019. Also, in March 2012, GAO reported that a now-deferred NNSA project to construct a new plutonium facility in Los Alamos, New Mexico, could cost as much as $5.8 billion, a nearly six-fold cost increase.

Chairman Murphy,
Ranking Member DeGette, and Members of the Subcommittee:

Thank you for the opportunity to discuss our work on the security, safety, and project management issues related to the nation's nuclear security enterprise. As you know, the National Nuclear Security Administration (NNSA), a separately organized agency within the Department of Energy (DOE), is responsible for managing nuclear weapon- and nonproliferation-related missions in research and development laboratories, production plants, and other facilities—known collectively as the nuclear security enterprise.[1] NNSA directs these national security missions, but work activities are largely carried out by management and operating (M&O) contractors at each site within the nuclear security enterprise. Working under M&O contracts, NNSA contractors apply their scientific, technical, and management expertise at NNSA's government-owned, contractor-operated sites.[2]

As we testified before this Subcommittee in September 2012, [3] questions have been raised about DOE's and NNSA's oversight of security, safety, and project and contract management for the nuclear security enterprise. For example, we first designated DOE's management of its contracts as an area at high risk of fraud, waste, abuse, and mismanagement in 1990 because of the department's record of inadequate management and oversight of its contractors. During the late 1990s, DOE experienced security problems at the nation's nuclear weapons laboratories and significant cost overruns on major projects (i.e., $750 million or more). According to a June 1999 report by the President's Foreign Intelligence Advisory Board, DOE's management of the nuclear weapons laboratories, while representing "science at its best," also embodied "security at its worst" because of "organizational disarray, managerial neglect, and a culture of arrogance." The advisory board urged Congress to create a new organization that, whether established as an independent agency or a semiautonomous agency within DOE, would have a clear mission, streamlined bureaucracy, and drastically simplified lines of authority and accountability. Responding to the advisory board's recommendations, Congress created NNSA under Title 32 of the National Defense Authorization Act for Fiscal Year 2000—the NNSA Act.[4] The NNSA Act established NNSA as a "separately organized agency" within DOE. The act established the position of DOE Under Secretary for Nuclear Security, who was also designated as the Administrator of NNSA. The Secretary of Energy and the Deputy Secretary of Energy were allowed to establish policy for NNSA and to give direction to NNSA through the Administrator; however, other DOE employees were prohibited from directing the activities of individual NNSA employees. DOE directives remain the primary means to establish, communicate, and institutionalize policies, requirements, responsibilities, and procedures for multiple departmental elements, including NNSA, but the act gives the NNSA Administrator the authority to establish NNSA-specific policies, unless disapproved by the Secretary of Energy. NNSA does this through the issuance of Policy Letters.[5]

NNSA's creation, however, has not yet had the desired effect of fully resolving long-standing management problems. For example, NNSA and DOE's Office of Environmental Management (EM) remain on our high-risk list.[6] Furthermore, we have frequently reported on security incidents and safety issues that have contributed to the temporary shutdown of facilities, such as at Los Alamos and Lawrence Livermore National Laboratories in 2004 and 2005,[7] as well as the safety and security initiatives that contractors at these laboratories put in place to help ensure improvement. More recently, at the Y-12 National Security Complex, in July 2012, three trespassers gained access to the protected security area directly adjacent to one of the nation's most critically important nuclear weapon-related facilities without being interrupted by the security measures in place. According to DOE's Inspector General, this security incident was unprecedented and represented multiple system failures including failures to maintain critical security equipment, respond properly to alarms, and understand security protocols.[8] Furthermore, the Inspector General found that contractor governance and federal oversight did not identify and correct early indications of these multiple system breakdowns.

DOE's management approach to security over the years has shifted in part due to concerns raised by some national laboratory, DOE, and NNSA officials. These officials believed that DOE's and NNSA's oversight of the laboratories' activities had become excessive and that the safety and security requirements for the laboratories are overly prescriptive and burdensome, which had resulted in a negative effect on the quality of science performed at these laboratories. Partly in response to these concerns, DOE and NNSA embarked on reforms in 2010 and 2011 that sought to streamline requirements and institute what has been called by the National Research Council, the DOE Inspector General, the DOE Office of Health and Safety Performance, and the Defense Nuclear Facilities Safety Board (Safety Board) a "hands-off, eyes-on" role for federal oversight. This approach placed more reliance on contractors' self-oversight through its contractor assurance systems to ensure such things as effective safety and security performance. 9 Building on this theme, in February 2012, the National Research Council found that "safety and security systems at the [NNSA] Laboratories have been strengthened to the point where they no longer need special attention."[10]

In this context, there have been calls to enhance NNSA's ability to operate independently of DOE. For example, the Defense Science Board proposed in 2006 that a completely independent nuclear weapons agency be created.[11] In January 2007, we reported that former senior DOE and NNSA officials with whom we spoke generally did not favor removing NNSA from DOE.[12] Furthermore, in a June 2012 report, we concluded that such a drastic change was unnecessary to produce an effective organization,[13] and we generally hold this view today. However, in the wake of the Y-12 security incident and persistent problems with major projects, there have been renewed calls to reexamine NNSA's organization. Most recently, the Fiscal Year 2013 National Defense Authorization Act created the Congressional Advisory Panel on the Governance of the Nuclear Security Enterprise to examine options and make recommendations for revising the governance structure, mission, and management of the nuclear security enterprise.

My testimony today discusses DOE's and NNSA's management of the nuclear security enterprise. It focuses on our reports issued from August 2000 to December 2012 on oversight of (1) security performance, (2) safety performance, and (3) project and contract management in the nuclear security enterprise. Detailed information about the scope and methodology can be found in our previously issued reports. We conducted the performance audit work that supports this statement in accordance with generally accepted government auditing standards. Those standards require that we plan and perform the audit to obtain sufficient, appropriate evidence to provide a reasonable basis for our findings and conclusions based on our audit objectives. We believe that the evidence obtained provides a reasonable basis for our findings and conclusions based on our audit objectives.

BACKGROUND

DOE is responsible for a diverse set of missions, including nuclear security, energy research, and environmental cleanup. These missions are managed by organizations within DOE and largely carried out by M&O contractors at various DOE sites. According to federal budget data, NNSA is one of the largest organizations in DOE, overseeing nuclear weapons, nuclear nonproliferation, and naval reactors missions at its sites. With an $11 billion budget in fiscal year 2012—nearly 40 percent of DOE's total budget—NNSA is responsible for providing the United States with safe, secure, and reliable nuclear weapons in the absence of underground nuclear testing and maintaining core competencies in nuclear weapons science, technology, and engineering. Ensuring a safe and reliable nuclear weapons stockpile is an extraordinarily complicated task and requires state-of-the-art experimental and computing facilities, as well as the skills of top scientists in the field. To its credit, NNSA consistently accomplishes this task, as evidenced by the successful assessment of the safety, reliability, and performance of each weapon type in the nuclear stockpile since its creation. To support these capabilities into the future, in 2011, the administration announced plans to request $88 billion from Congress over the next decade to operate and modernize the nuclear security enterprise.

As discussed earlier, work activities to support NNSA's national security missions are largely carried out by M&O contractors. This arrangement has historical roots. Since the Manhattan Project produced the first atomic bomb during World War II, NNSA, DOE, and predecessor agencies have depended on the expertise of private firms, universities, and others to carry out research and development work and efficiently operate the facilities necessary for the nation's nuclear defense. Currently, DOE spends 90 percent of its annual budget on M&O contracts, making it the largest non-Department of Defense contracting agency in the government.

DOE generally regulates the safety of its own nuclear facilities and operations at its sites. In contrast, the Nuclear Regulatory Commission (NRC) generally regulates commercial nuclear facilities, and the Occupational Safety and Health Administration (OSHA) generally regulates worker safety at commercial industrial facilities.[14] However, because of the dangerous nature of work conducted at many sites within the nuclear security enterprise—handling nuclear material such as plutonium, manufacturing high explosives, and various industrial operations that use hazardous chemicals—oversight of the nuclear security enterprise is multifaceted. First, DOE policy states that its M&O contractors are expected to develop and implement an assurance system, or system of management controls, that helps ensure the department's program missions and activities are executed in an effective, efficient, and safe manner.[15] Through these assurance systems, contractors are required to perform self-assessments as well as identify and correct negative performance trends. Second, NNSA site offices, which are collocated with NNSA sites, oversee the performance of M&O contractors. Site office oversight includes communicating performance expectations to the contractor, reviewing the contractor's assurance system, and conducting contractor performance evaluations. Third, DOE's Office of Health, Safety, and Security—especially its Office of Independent Oversight—conducts periodic appraisals to determine if NNSA officials and contractors are complying with safety and security requirements.[16] Fourth, NNSA receives safety assessments and recommendations from other organizations, most prominently the Safety Board—an independent executive branch agency created by Congress to assess safety conditions and operations at DOE's defense nuclear facilities.[17] To address public health and safety issues, the Safety Board is authorized to make recommendations to the Secretary of Energy, who may then accept or reject, in whole or in part, the recommendations. If the Secretary of Energy accepts the recommendations, the Secretary must prepare an implementation plan.

DOE and some of its contractors have viewed this multifaceted oversight to be overly burdensome. To address this issue, in March 2010 the Deputy Secretary of Energy announced a reform effort to revise DOE's safety and security directives and modify the department's oversight approach to "provide contractors with the flexibility to tailor and implement safety and security programs without excessive federal oversight or overly prescriptive departmental requirements." In the memorandum announcing this effort, the

Deputy Secretary noted that burdensome safety requirements were affecting the productivity of work at DOE's sites and that reducing this burden on contractors would lead to measurable productivity improvement. The Deputy Secretary noted that DOE's Office of Health, Safety and Security in 2009 had begun reforming its approach to enforcement and oversight. Similar to, but independent of DOE's safety and security reform effort, in February 2011, NNSA initiated its "governance transformation" project, which involved revising the agency's business model to, among other things, place more reliance on contractors' self-oversight through its contractor assurance system to ensure such things as effective safety and security performance. NNSA's non-nuclear Kansas City Plant completed implementation of this new business model, and other NNSA sites—such as the Nevada National Security Site and the Y-12 National Security Complex—were in the process of implementing it, too, when the Y-12 security incident occurred.

DOE's and NNSA's Oversight of Security Performance Continues to Face Challenges

In response to the Y-12 security breach, multiple investigations and reviews of the incident were performed by NNSA, the DOE Office of Inspector General, and the DOE Office of Independent Oversight. These reviews identified numerous problems with NNSA's and its contractors' performance, including: physical security systems, such as alarms; protective force (i.e., NNSA's heavily armed, contractor guard forces) training and response; failures to correct numerous known problems; and weaknesses in contract and resource management. In addition, at the request of the Secretary of Energy, an independent panel, composed of three former executives from Federal agencies and the private sector, and a NNSA Security Task Force found broader and systemic security issues across the nuclear security enterprise. The Secretary's panel in December 2012 analyzed various models for providing security at DOE and NNSA sites but generally found that improvements to the security culture, management, and oversight were necessary, in addition to having an effective organizational structure. In addition, the leader of the NNSA Security Task Force testified before the House Armed Services Committee in February 2013 about significant deficiencies in NNSA's entire security organization, oversight, and culture.

In response to the Y-12 security incident and these findings, DOE and NNSA took a number of immediate actions, including repairing security equipment, reassigning key security personnel, and firing the Y-12 protective force contractor. In February 2013, the Acting NNSA Administrator committed to implementing a three-tiered oversight process involving contractor self-assessment, NNSA evaluation of site performance, and independent oversight by DOE's Office of Independent Oversight. The Acting Administrator testified before the House Armed Services Committee that she believed that such actions will help instill a culture that embraces security as an essential element of NNSA's missions.

In assessing DOE's actions to address the security breakdowns at Y-12, a central question will be whether these latest actions taken will produce sustained improvements in security at Y-12 and across the nuclear security enterprise. As we and others have reported, DOE has a long history of security breakdowns and an equally long history of instituting responses and remedies to "fix" these problems. For example, in examining the Y-12 security incident, NNSA's former Acting Chief of Defense Nuclear Security and the leader of the NNSA's Security Task Force testified in February 2013 about problems with NNSA's federal security organization including poorly defined roles and responsibilities for its headquarters and field security organizations, inadequate oversight and assessments of site security activities, and issues with overseeing contractor actions and implementing improvements. As noted in table 1, 10 years ago we reported on very similar problems, and since that time DOE has undertaken numerous security initiatives to address them. We have not evaluated these recent initiatives but we have ongoing work to evaluate them as part of our review on security reform for the Subcommittee, which we will complete later this year.

It is also important to note that NNSA's long-standing security problems are not limited to Y-12. DOE's and NNSA's work with nuclear materials such as plutonium and highly enriched uranium, nuclear weapons and their components, and large amounts of classified data require extremely high security, however, as we and DOE have reported, NNSA and DOE have a long history of poor security performance across the nuclear security enterprise, most notably at Los Alamos and Livermore national laboratories, as well as ongoing struggles to sustain security improvements, including information security.[18]

Table 1. Comparison of GAO 2003 Findings Regarding NNSA's Federal Security Organization with NNSA Security Task Force's February 2013 Findings

	GAO May 2003 Findings	NNSA Security Task Force February 2013 Findings
Defining clear roles and responsibilities	NNSA has not fully defined clear roles and responsibilities for its headquarters and site operations.	NNSA security line management authority is ill-defined. There are overlapping lines of authority and a mixing of staff and line functions.
Assessing sites' security activities	There are inconsistencies among NNSA sites on how they assess contractors' security activities. Consequently, NNSA cannot be assured that all facilities are subject to comprehensive annual assessments as required by DOE policy.	NNSA does not have an adequate security performance assessment process or capability. NNSA has come to rely overwhelmingly on federal staff reviewing contractor-provided data, rather than effectively assessing performance itself.
Overseeing contractors' actions and implementing long-term improvements	NNSA contractors do not consistently conduct required analyses in preparing security corrective program action plans. Security performance at sites may not be maximized because corrective security program actions are developed without fully considering root causes, risks posed, or costs and benefits of taking corrective action.	NNSA has attempted to correct some identified issues over the years, but it has not adequately emphasized security mission performance. RecentDOE and NNSA reforms have deemphasized performance verification by federal staff, resulting in a weakened federal security assessment program.
Allocating staff	NNSA has shortfalls at its site offices in number and expertise of staff, which could make it more difficult for site offices to effectively oversee security activities.	The NNSA federal security function is not properly organized or staffed.

Sources: GAO, *Nuclear Security: NNSA Needs to Better Manage Its Safeguards and Security Program*, GAO-03-471 (Washington, D.C.: May 30, 2003); and Hearing on Nuclear Security: Actions, Accountability, and Reform, Before the Subcommittee on Strategic Forces, House Armed Services Committee, 113th Cong. (Feb. 28, 2013) (Statement of Brigadier General Sandra E. Finan, Commander, Air Force Nuclear Weapons Center. Testimony based on Previous Position as Acting Chief of Defense Nuclear Security, NNSA).

Los Alamos National Laboratory

As we noted in our September 2012 testimony,[19] Los Alamos National Laboratory (Los Alamos) experienced a number of high-profile security incidents in the previous decade that were subject to congressional hearings, including some held by this Subcommittee. Many of these incidents focused on Los Alamos's inability to account for and control its classified resources. These incidents include the transfer or removal of classified information from authorized work areas or the laboratory itself, the temporary loss of two hard drives containing nuclear weapon design information, and difficulties in accounting for classified removable electronic media. In addition to these well-publicized incidents, security evaluations through 2007 identified other persistent, systemic security problems at Los Alamos. These problems included weaknesses in controlling and protecting classified resources, inadequate controls over special nuclear material, inadequate self-assessment activities, and weaknesses in the process that Los Alamos uses to ensure it corrects identified security deficiencies. Partly as a result of these findings, as we reported in 2008,[20] Los Alamos underwent a 10 month shut-down of operations in 2004 and experienced a change in contractors in 2005. Moreover, the Secretary of Energy issued a compliance order in 2007 requiring Los Alamos to implement specific corrective actions to, among other things, address long-standing deficiencies in its classified information programs. We reported in January 2008 and testified before this Subcommittee in September 2008 that Los Alamos had experienced a period of improved security performance but that it was too early to determine whether NNSA and Los Alamos could sustain this level of improvement.[21]

Lawrence Livermore National Laboratory

In March 2009, we reported on numerous and wide-ranging security deficiencies at Lawrence Livermore National Laboratory (Livermore), particularly in the ability of Livermore's protective forces to ensure the protection of special nuclear material and the laboratory's protection and control of classified matter.[22] We also identified Livermore's physical security systems, such as alarms and sensors, and its security program planning and assurance activities, as areas needing improvement. Weaknesses in Livermore's contractor self-assessment program and the Livermore Site Office's oversight of the contractor contributed to these security deficiencies at the laboratory. According to one DOE Office of Independent Oversight official, both programs were "broken" and missed even the "low-hanging fruit." The laboratory took corrective action to address these deficiencies, but we noted that better oversight was needed to ensure that security improvements were fully implemented and sustained. In September 2012, NNSA and Livermore completed efforts to move the site's most sensitive nuclear material to other sites, thereby easing the site's security requirements.

Information Security

We also have reported extensively on NNSA's challenges in maintaining effective and secure information security systems, particularly at Los Alamos. For example, in June 2008, we reported that significant information security problems at Los Alamos had received insufficient attention.[23] The laboratory had over two dozen initiatives under way that were principally aimed at reducing, consolidating, and better protecting classified resources. However, the laboratory had not implemented complete security solutions to address either the problems of classified parts storage in unapproved storage containers or weaknesses in its process for ensuring that actions taken to correct security deficiencies were completed. In addition, in October 2009 we reported that Los Alamos needed to better protect its classified network.[24] Specifically, we found significant weaknesses remained in protecting the confidentiality, integrity, and availability of information stored on and transmitted over its classified computer network. Moreover, we found the laboratory's decentralized approach to information security program management has led to inconsistent implementation of policy.

DOE and NNSA's Oversight of Safety Performance Continues to Face Challenges

DOE and NNSA have experienced significant safety problems at their sites, and recent efforts to reform safety protocols and processes have not demonstrated sustained improvements. As we testified in September 2012 before this Subcommittee,[25] long-standing DOE and NNSA management weaknesses have contributed to persistent safety problems at NNSA's national laboratories. For example, in October 2007, we reported that nearly 60 serious accidents or near misses had occurred at NNSA's national laboratories since 2000.[26] These accidents included worker exposure to radiation, inhalation of toxic vapors, and electrical shocks. Although no one was killed, many of these accidents caused serious harm to workers or damage to facilities. As we also reported, at Los Alamos in July 2004, an undergraduate student who was not wearing required eye protection was partially blinded in a laser accident. Our review of nearly 100 safety studies—including accident investigations and independent assessments by the Safety Board and others issued since 2000—found that the contributing factors to these safety problems generally fell into three key categories: (1) relatively lax laboratory attitudes toward safety procedures, (2) laboratory inadequacies in identifying and addressing safety problems with appropriate corrective actions, and (3) inadequate oversight by NNSA site offices.[27] DOE's Office of Inspector General has also raised concerns about safety oversight by NNSA's site offices. Specifically, the Inspector General reported in June 2011 that NNSA's Livermore Site Office was not sufficiently overseeing its contractor to ensure that corrective actions were fully and effectively implemented for a program designed to limit worker exposure to beryllium, a hazardous metal essential for nuclear operations.[28]

DOE has undertaken a number of reforms to address persistent safety concerns. In March 2010, the Deputy Secretary of Energy announced a reform effort to revise DOE's safety and security directives. The reform effort was aimed at modifying the department's oversight approach to "provide contractors with the flexibility to tailor and implement safety and security programs without excessive federal oversight or overly prescriptive departmental requirements." As we reported to this Subcommittee in April 2012,[29] this reform effort reduced the number of safety related directives from 80 to 42 by eliminating or combining requirements the department determined were unclear, duplicative, or too prescriptive and by encouraging the use of industry standards. However, as we noted in September 2012 before this Subcommittee, DOE's safety reforms did not fully address safety concerns that we, as well as others, have identified in the areas of quality assurance, safety culture, and federal oversight and, in fact, these reforms may have actually weakened independent oversight. We stated, for example, that while DOE policy notes that independent oversight is integral to help ensure the effectiveness of safety performance, DOE's Office of Independent Oversight staff must now coordinate their assessment activities with NNSA site office management to maximize the use of resources. This arrangement raised our concern about whether Office of Independent Oversight staff would be sufficiently independent from site office management. In our April 2012 report, we recommended, among other things, that DOE develop a detailed reform plan and clearly define the oversight roles and responsibilities of DOE's Office of Independent Oversight staff to ensure that their work is sufficiently independent from the activities of DOE site office and contractor staff. DOE has taken steps to respond to these recommendations, including developing a plan aimed at improving safety management and drafting a memo from the Secretary of Energy reconfirming the department's commitment to independent oversight of safety and security.

However, since our September 2012 testimony, 30 concerns continue to be raised about safety performance and oversight at several NNSA sites, which indicate DOE's safety reforms have not brought about a sustained change in safety practices. The following are examples:

- A November 2012 report by DOE's Office of Independent Oversight raised concerns about safety culture issues at NNSA's Pantex Plant. Among the concerns were reluctance by workers to raise safety problems for fear of retaliation and a perception that cost took priority over safety.
- At an October 2012 public hearing in Knoxville, Tennessee, the Safety Board noted that safety controls to prevent or mitigate consequences from accidents had not been fully incorporated into the design of a new uranium processing facility at Y-12. The Safety Board noted the facility's safety basis—a technical analysis that identifies potential accidents and hazards associated with a facility's operations and outlines controls to mitigate or prevent their impact on workers and the public—did not adequately address controls to protect workers or the public in the case of an earthquake or small fires, and did not adequately calculate reasonably conservative radiation exposure consequences that could lead to putting greater safety into the facility's design. The Safety Board further noted that these deficiencies raise the potential for significant impacts on public and worker safety.
- A January 2013 Office of Independent Oversight report reviewing the Los Alamos Site Office assessment of the contractor corrective action system found that the contractor had not implemented effective corrective actions for identified safety system problems. This report noted that the site office concluded that more than half of the 62 safety system items needing corrective action had been closed without adequate action or sufficient documentation. Moreover, in October

2012, NNSA issued a Preliminary Notice of Violation to a Los Alamos contractor for repeated electrical safety problems. NNSA's notice stated that insufficient oversight of subcontractor work by the contractor safety staff was among the contributing factors. NNSA fined the contractor $262,500.

DOE and NNSA Have Made Progress but Further Improvements Needed on Project and Contract Management

A basic tenet of effective management is the ability to complete projects on time and within budget. DOE has taken a number of actions to improve management of projects, including those overseen by NNSA. For example, DOE has updated project and contract management policies and guidance in an effort to improve the reliability of project cost estimates, better assess project risks, and better ensure project reviews that are timely and useful and identify problems early. In addition, in December 2010, the Deputy Secretary of Energy convened a DOE Contract and Project Management Summit to discuss strategies for additional improvement in contract and project management. The participants identified barriers to improved performance and reported in April 2012 on the status of initiatives to address these barriers. DOE has continued to release guides for implementing its revised order for Program and Project Management for the Acquisition of Capital Assets (DOE O 413.3B), such as for cost estimating, using earned value management, and forming project teams. Further, DOE has taken steps to enhance project management and oversight by requiring peer reviews and independent cost estimates for projects with values of more than $100 million and by improving the accuracy and consistency of data in its central repository for project data.

DOE has made progress in managing nonmajor projects—those costing less than $750—million and in recognition of this progress, we narrowed the focus of our high-risk designation to major contracts and projects. Specifically, as we noted in our October 2012 report on DOE's EM cleanup projects funded by the American Recovery and Reinvestment Act, at the time of our analysis, 78 of 112 projects had been completed.[31] Of those completed projects, 92 percent met the performance standard of completing project work scope without exceeding the cost target by more than 10 percent, according to EM data. However, we made four recommendations to DOE in this report aimed at improving how EM manages and documents projects, particularly with respect to establishing key performance parameters such as project scope targets and baselines for cost and schedule. DOE concurred with all of our recommendations, recognizing that improvements could be made and that lessons learned from these projects can be applied to EM's broader portfolio of projects and activities. In addition, in December 2012, we reported that EM and NNSA were making some progress in managing the 71 nonmajor construction and cleanup projects that we reviewed and are expected to cost an estimated $10.1 billion in total.[32] For example, we identified some NNSA and EM nonmajor projects that used sound project management practices, such as the application of effective acquisition strategies, to help ensure the successful completion of these projects. We also recommended that NNSA and EM clearly define, document, and track the scope, cost, and completion date targets for each of their nonmajor projects and that EM clearly identify critical occupations and skills in its workforce plans. NNSA and EM agreed with these recommendations.

Notwithstanding these positive developments for nonmajor projects, major projects (i.e., those $750 million or more) continue to pose a challenge for DOE and NNSA. Since 1990 when we placed contract and project management on the high-risk list, we have reported on problems that principally involve ineffective oversight and poor management of contractors. Our recent work, as well as reporting by DOE, indicates that these problems continue. Examples are as follows:

- In December 2012, we reported that the estimated cost to construct the Waste Treatment and Immobilization Plant in Washington State had almost tripled to $12.3 billion since the project's inception in 2000, and the scheduled completion date had slipped by nearly a decade to 2019.[33] Moreover, we found that DOE's incentives and management controls were inadequate for ensuring effective project management and we also found instances where DOE prematurely rewarded the contractor for resolving technical issues and completing work.
- In March 2012, we reported that NNSA's project to construct a new plutonium facility, the Chemistry and Metallurgy Research Replacement Nuclear Facility, at Los Alamos could cost as much as $5.8 billion, a nearly six-fold increase from its original estimate. 34 While the facility may be large enough to support nuclear weapon stockpile requirements, our March 2012 report found that it is unclear

if the facility will be large enough to accommodate DOE's non-weapon activities that involve plutonium—such as nonproliferation, nuclear forensics, and nuclear counterterrorism programs—because the department has not comprehensively studied its long-term research and storage needs.

- In November 2010, we reported that NNSA's plans to construct a modern Uranium Processing Facility at its Y-12 National Security Complex in Oak Ridge, Tennessee, had experienced significant cost increases.[35] More recently, in September 2011, NNSA estimated that the facility would cost from $4.2 billion to $6.5 billion to construct—a nearly seven-fold cost increase from the original estimate.
- In April 2010, we reported that weak management by DOE and NNSA had allowed the cost, schedule, and scope of ignition-related activities at the National Ignition Facility to increase substantially.[36] We reported that, since 2005, ignition-related costs have increased by around 25 percent—from $1.6 billion in 2005 to over $2 billion in 2010—and that the planned completion date for these activities had slipped from the end of fiscal year 2011 to the end of fiscal year 2012 or beyond. Ten years earlier, in August 2000, we had reported that poor management and oversight of the National Ignition Facility construction project at Lawrence Livermore National Laboratory had increased the facility's cost by $1 billion and delayed its scheduled completion date by 6 years.[37]
- In March 2010, we reported that NNSA's Mixed-Oxide Fuel Fabrication Facility currently being constructed at DOE's Savannah River Site in South Carolina had experienced delays, but project officials said that they expected to recover from these delays by the end of 2010 and planned for the start of operations on schedule in 2016. In addition, after spending about $730 million on design, NNSA has cancelled the pit disassembly and conversion facility and is now planning to use existing facilities at DOE's Savannah River and Los Alamos sites and will add equipment to the mixed oxide facility. NNSA is working on a cost and schedule estimate for the use of these existing facilities and for adding the additional equipment.

We have also issued several reports on the technical issues, cost increases, and schedule delays associated with NNSA's efforts to extend, through refurbishment, the operational lives of nuclear weapons in the stockpile. For example, in March 2009, we reported that NNSA and the Department of

Defense had not effectively managed cost, schedule, and technical risks for the B61 nuclear bomb and the W76 nuclear warhead refurbishments.[38] For the B61 life extension program, NNSA was only able to stay on schedule by significantly reducing the number of weapons undergoing refurbishment and abandoning some refurbishment objectives. Earlier, in December 2000, we similarly had reported that refurbishment of the W87 strategic warhead had experienced significant design and production problems that increased its refurbishment costs by over $300 million and caused schedule delays of about 2 years.[39]

In conclusion, the actions that DOE and NNSA have taken to address weaknesses in oversight of security, safety, and contract and project management are very important, but problems persist. While we have noted progress in the area of project management, we also observe that NNSA and DOE EM have not begun a new major project since taking these actions. The Y-12 security incident was an unprecedented event for the nuclear security enterprise and perhaps indicates that NNSA's organizational culture, over a decade after the agency was created to address security issues, still has not embraced security as an essential element of its missions. In terms of safety, DOE has recently taken the initiative to examine the safety culture at its sites. We believe, as do other organizations, including the DOE Inspector General and Safety Board, that a "hands off, eyes on" oversight approach for security, safety and contract and project management is insufficient and unwarranted until the department can demonstrate sustained improvement in all three areas. We will continue to monitor DOE's and NNSA's implementation of actions to resolve its safety, security, and contract and project management difficulties and to assess the impact of these actions.

Chairman Murphy, Ranking Member DeGette, and Members of the Subcommittee, this completes my prepared statement. I would be pleased to respond to any questions you may have at this time.

End Notes

[1] Specifically, NNSA manages three national nuclear weapon design laboratories— Lawrence Livermore National Laboratory in California, Los Alamos National Laboratory in New Mexico, and Sandia National Laboratories in New Mexico and California. It also manages four nuclear weapon production plants—the Pantex Plant in Texas, the Y-12 National Security Complex in Tennessee, the Kansas City Plant in Missouri, and the Tritium Extraction Facility at DOE's Savannah River Site in South Carolina. NNSA also manages the Nevada National Security Site, formerly known as the Nevada Test Site.

[2] M&O contracts are agreements under which the federal government contracts for the operation, maintenance, or support, on its behalf, of a government-owned or -controlled research, development, special production, or testing establishment wholly or principally devoted to one or more of the major programs of the contracting federal agency. Federal Acquisition Regulation, 48 C.F.R. § 17.601 (2012).

[3] GAO, *Modernizing the Nuclear Security Enterprise: Observations on the National Nuclear Security Administration's Oversight of Safety, Security, and Project Management*, GAO-12-912T (Washington, D.C.: Sept. 12, 2012).

[4] Pub. L. No. 106-65, 113 Stat. 512, 953 (1999).

[5] NNSA, *Policy Letters: NNSA Policies, Supplemental Directives, and Business Operating Procedures,* NA SD 251.1 (Washington, D.C.: July 5, 2011).

[6] GAO, *High-Risk Series: An Update*, GAO-13-283 (Washington, D.C.: Feb. 2013). We have shifted our focus concerning the high-risk area to major DOE-EM and NNSA major projects (i.e., those $750 million or more).

[7] For additional information on the 2004 temporary shutdown of facilities at Los Alamos, see GAO, *Stand-Down of Los Alamos National Laboratory: Total Costs Uncertain; Almost All Mission-Critical Programs Were Affected but Have Recovered*, GAO-06-83 (Washington, D.C.: Nov. 18, 2005).

[8] DOE Office of Inspector General, *Inquiry into the Security Breach at the National Nuclear Security Administration's Y-12 National Security Complex*, DOE/IG-0868 (August 2012).

[9] These systems include management controls that help ensure the department's program missions and activities are executed in an effective, efficient, and safe manner. We are currently evaluating the implementation of contractor assurance systems at NNSA sites and NNSA's oversight of these systems.

[10] National Research Council, *Managing for High-Quality Science and Engineering at the NNSA National Security Laboratories*, (Washington, D.C.: Feb. 15, 2012).

[11] The Defense Science Board provides the Department of Defense with independent advice and recommendations on matters relating to the department's scientific and technical enterprise. See Defense Science Board Task Force, *Nuclear Capabilities* (Washington, D.C.: Dec. 2006).

[12] GAO, *National Nuclear Security Administration: Additional Actions Needed to Improve Management of the Nation's Nuclear Programs*, GAO-07-36 (Washington, D.C.: Jan.19, 2007).

[13] GAO, *Modernizing the Nuclear Security Enterprise: Observations on the Organization and Management of the National Nuclear Security Administration*, GAO-12-867T (Washington, D.C.: June 27, 2012).

[14] DOE regulates the safety of most of its own sites with nuclear operations; NRC regulates several DOE nuclear facilities, and OSHA regulates occupational safety at DOE sites that have no nuclear function.

[15] DOE, *Department of Energy Oversight Policy*, DOE P 226.1B (Washington, D.C.: Apr. 25, 2011). Contractor assurance systems are to cover the following operational aspects: (1) environment, safety, and health; (2) safeguards and security; (3) emergency management; and (4) cyber security.

[16] DOE reorganized offices within its Office of Health, Safety, and Security. The Office of Independent Oversight merged with the Office of Enforcement and was renamed the Office of Enforcement and Oversight. For the purposes of this report, we refer to it as the Office of Independent Oversight.

[17] The Safety Board provides oversight for all NNSA sites except the Kansas City Plant, which manufactures non-nuclear components.

[18] We note that over the past decade, the DOE Inspector General and Office of Independent Oversight periodically identified serious security issues at almost all of NNSA's sites, including Sandia National Laboratories; the Nevada National Security Site, Pantex, and prior to the July 2012 security incident, the Y-12 National Security Complex.

[19] GAO-12-912T.

[20] GAO, *Los Alamos National Laboratory: Long-Term Strategies Needed to Improve Security and Management Oversight*, GAO-08-694 (Washington, D.C.: June 13, 2008).

[21] GAO, *Los Alamos National Laboratory: Information on Security of Classified Data, Nuclear Material Controls, Nuclear and Worker Safety, and Project Management Weaknesses*, GAO-08-173R (Washington D.C.: Jan. 10, 2008), and GAO, *Nuclear Security: Los Alamos National Laboratory Faces Challenges in Sustaining Physical and Cyber Security Improvements,* GAO-08-1180T (Washington, D.C.: Sep. 25, 2008).

[22] GAO, *Nuclear Security: Better Oversight Needed to Ensure That Security Improvements at Lawrence Livermore National Laboratory Are Fully Implemented and Sustained*, GAO-09-321 (Washington, D.C.: Mar. 16, 2009).

[23] GAO, *Los Alamos National Laboratory: Long-Term Strategies Needed to Improve Security and Management Oversight*, GAO-08-694 (Washington, D.C.: June 13, 2008).

[24] GAO, *Information Security: Actions Needed to Better Manage, Protect, and Sustain Improvements to Los Alamos National Laboratory's Classified Computer Network*, GAO-10-28 (Washington, D.C.: Oct. 14, 2009).

[25] GAO-12-912T.

[26] GAO, *Nuclear and Worker Safety: Actions Needed to Determine the Effectiveness of Safety Improvement Efforts at NNSA's Weapons Laboratories*, GAO-08-73 (Washington, D.C.: Oct. 31, 2007).

[27] GAO-08-73.

[28] DOE Office of Inspector General, *Implementation of Beryllium Controls at Lawrence Livermore National Laboratory,* DOE/IG-0851 (Washington, D.C.: June 17, 2011).

[29] GAO, *Nuclear Safety: DOE Needs to Determine the Costs and Benefits of Its Safety Reform Effort*, GAO-12-347 (Washington, D.C.: Apr. 20, 2012).

[30] GAO-12-912T.

[31] GAO, *Recovery Act: Most DOE Projects Are Complete, but Project Management Guidance Could Be Strengthened*, GAO-13-23 (Washington D.C.: Oct. 15, 2012).

[32] GAO, *Department of Energy: Better Information Needed to Determine If Nonmajor Projects Meet Performance Targets*, GAO-13-129 (Washington, D.C.: Dec. 19, 2012).

[33] GAO, *Hanford Waste Treatment Plant: DOE Needs to Take Action to Resolve Technical and Management Challenges*, GAO-13-38 (Washington, D.C.: Dec. 19, 2012).

[34] GAO, *Modernizing the Nuclear Security Enterprise: New Plutonium Research Facility at Los Alamos May Not Meet All Mission Needs*, GAO-12-337 (Washington, D.C.: Mar. 26, 2012).

[35] GAO, *Nuclear Weapons: National Nuclear Security Administration's Plans for Its Uranium Processing Facility Should Better Reflect Funding Estimates and Technology Readiness*, GAO-11-103 (Washington, D.C.: Nov. 19, 2010).

[36] Ignition-related activities consist of the efforts separate from the facility's construction that have been undertaken to prepare for the first attempt at ignition—the extremely intense pressures and temperatures that simulate on a small scale the thermonuclear conditions created in nuclear explosions. See GAO, *Nuclear Weapons: Actions Needed to Address*

Scientific and Technical Challenges and Management Weaknesses at the National Ignition Facility, GAO-10-488 (Washington, D.C.: Apr. 8, 2010).

[37] GAO, *National Ignition Facility: Management and Oversight Failures Caused Major Cost Overruns and Schedule Delays*, GAO/RCED-00-271 (Washington, D.C.: Aug. 8, 2000).

[38] GAO, *Nuclear Weapons: NNSA and DOD Need to More Effectively Manage the Stockpile Life Extension Program*, GAO-09-385 (Washington, D.C.: Mar. 2, 2009).

[39] GAO, *Nuclear Weapons: Improved Management Needed to Implement Stockpile Stewardship Program Effectively*, GAO-01-48 (Washington, D.C.: Dec. 14, 2000).

INDEX

D

E

F

G

H

I

O

P

Q

R